Springer Undergraduate Mathematics Series

SUMS Readings

SUMS Readings is a collection of books that provides students with opportunities to deepen understanding and broaden horizons. Aimed mainly at undergraduates, the series is intended for books that do not fit the classical textbook format, from leisurely-yet-rigorous introductions to topics of wide interest, to presentations of specialised topics that are not commonly taught. Its books may be read in parallel with undergraduate studies, as supplementary reading for specific courses, background reading for undergraduate projects, or out of sheer intellectual curiosity. The emphasis of the series is on novelty, accessibility and clarity of exposition, as well as self-study with easy-to-follow examples and solved exercises.

R. Michael Howe

An Invitation
to Representation Theory

Polynomial Representations
of the Symmetric Group

 Springer

R. Michael Howe
Department of Mathematics
University of Wisconsin–Eau Claire
Eau Claire, WI, USA

ISSN 1615-2085   ISSN 2197-4144   (electronic)
Springer Undergraduate Mathematics Series
ISSN 2730-5813   ISSN 2730-5821   (electronic)
SUMS Readings
ISBN 978-3-030-98024-5   ISBN 978-3-030-98025-2   (eBook)
https://doi.org/10.1007/978-3-030-98025-2

Mathematics Subject Classification: 20-01, 20B05, 20B30, 20B35, 20Cxx, 20C05, 20C30

This Springer imprint is published by the registered company Springer Nature Switzerland AG
The registered company address is: Gewerbestrasse 11, 6330 Cham, Switzerland

To my family, friends, colleagues, and students.
I'm only writing one book, so...

Preface

A good stack of examples, as large as possible, is indispensable for a thorough understanding of any concept, and when I want to learn something new, I make it my first job to build one.

Paul Halmos

Over the years I have had the opportunity to work with undergraduate mathematics students on various research and independent study projects, and I've found "representation theory of the symmetric group" to be an excellent vehicle to introduce them to more advanced mathematical concepts relatively early in their undergraduate careers. The basic idea is accessible to anyone with a solid background in linear algebra, and new content can be introduced as needed. Of course, it's also a beautiful subject in its own right that has many important applications.

The concept of the "symmetric group," that is, permutations acting on a set, along with the operation of composition and the existence of inverse permutations, is easy for the beginner to get their mind around, and serves as one of the first non-trivial examples of a non-Abelian group. Students are comfortable with polynomials, and the action of permutations on polynomials is straightforward and easy to understand.

Examples of such central concepts as "invariant subspaces," "irreducible subspaces," "isotypic subspaces," and "intertwining maps" can be explicitly constructed, and the further notions of "permutation representations" and "Young permutation modules" arise naturally. This setting provides examples that aid the intuition when moving on to more sophisticated territory, and much of the theory of the representations of so-called Lie groups is built on that of the symmetric group. On a somewhat historical note, some of the early study of the symmetric group and representation theory was by way of permuting the arguments of functions, called *substitutional analysis*. (See [R]).

There are LOTS of excellent books on the subject of representation theory, but they are essentially written at the graduate-student level (at least), and usually require familiarity with topics (topological spaces, tensor products, and multi-linear algebra) beyond the experience of, say, a typical second-year mathematics undergraduate. The intended audience for this book will have completed a solid course in linear algebra, and so should be familiar with abstract vector spaces and the basic techniques of mathematical proofs. The reader may also be taking a modern (or "abstract") algebra course concurrently, but I assume no knowledge of modern algebra. Readers certainly don't need to know everything from undergraduate algebra, or even group theory, and I introduce the necessary concepts as the need arises. I recommend that the novice reader have a good linear algebra textbook, as well as a modern algebra textbook, at hand for reference. I caution that navigating multiple such references can be counter-productive, if for no other reason than differing notation. I also highly recommend working with a colleague and/or, if available, someone with more knowledge of the subject. Additionally, facility with a computer algebra system (CAS) such as Maple or Mathematica can ease some of the linear algebra computations.

Readers who are more mathematically advanced, but new to the subject, may find the discussion and the examples useful, but I emphasize that this book is written for novice (but motivated) mathematicians. Many of the explanations and proofs include more detail than would typically be included in a text for more "mathematically mature" readers, with the solutions to the exercises being particularly, if not annoyingly, detailed. Given a choice between saying too much or too little, I have erred on the side of saying too much. More advanced readers should be able to skim through the excessive details with minimal irritation.

I make no claim of originality, other than, perhaps, the level of the intended audience and the number of exercises. The material here is taken from a variety of sources, chosen and adapted so that it is accessible to novice mathematicians. I have tried to cite most of the sources where applicable, although I make no claim as to completeness of citation. I also make no attempt at efficiency. Some results and remarks are repeated as needed for clarity or precision, and for the benefit of the "grasshopper reader" who may jump in and around the text. The subject suffers from flourishing terminology, and I occasionally use terms interchangeably.

Throughout I endeavor to use the active voice. The term "we" can mean the author and the reader, as in "we see that...." Often the term "we" can be interpreted as "the mathematical community," as in "we show that ...," since the ideas have been developed and exposed by dozens, if not hundreds, of mathematicians before us.

I would like to thank Chal Benson, Chris Davis, Carolyn Otto, and Gail Ratcliff (as well as the anonymous referees) for providing detailed and useful comments on early versions of this book. I would also like to express my appreciation to the scores of undergraduate students who participated in research and independent study projects with me, at least a dozen of whom have gone on to earn PhDs in mathematics.

Eau Claire, WI, USA   R. Michael Howe

Introduction

At this introductory level, the theory of groups and their representations lies mostly within the mathematical realm of algebra, so let's first review what this means. An *algebraic structure* consists of an underlying set A, along with some ways of combining the elements of A to obtain another element in A by using *operations* that are required to satisfy certain *axioms*.

There are lots of algebraic structures: vector spaces, semigroups, rings, lattices; the list goes on. There are also other types of mathematical structures such as topological and analytic structures as well as hybrid structures such as topological groups and Hilbert spaces. The main concern of this book will be the algebraic structure of groups and modules. Groups are ubiquitous in mathematics and make precise the notion of symmetry.

Group representations are a type of module, and representation theory translates many group-theoretic problems into problems in the (usually easier) landscape of linear algebra. The theory has wide-ranging applications, from quantum mechanics [W] to voting theory [D] and to other areas of mathematics (algebraic geometry, invariant theory, multivariate statistics, combinatorics). It is the opinion of some science writers that "representation theory has served as a key ingredient in many of the most important discoveries in mathematics." [H].

One algebraic structure that readers are assumed to be familiar with is that of a *vector space*, with the operations of scalar multiplication and vector addition that satisfy the various associative, commutative, and distributive properties, etc. Some readers may wish to review the properties of vector spaces.

The scalars in a vector space are from a *field*, another algebraic structure with which readers should be familiar. The most common examples of fields are the rational numbers $\mathbb{Q}$, the real numbers $\mathbb{R}$, and the complex numbers $\mathbb{C}$, along with the operations of addition and multiplication that satisfy the various axioms of associativity, existence of identities and inverses, etc. Again, some readers may wish to review the properties of fields.

The *characteristic* of a field $\mathbb{F}$ is defined as the smallest natural number n such that[1]

$$na := \underbrace{a + a + \cdots + a}_{n \ summands} = 0$$

for any $a \in \mathbb{F}$. If no such n exists, we say $\mathbb{F}$ has *characteristic zero*, as is the case with the familiar fields $\mathbb{Q}$, $\mathbb{R}$, and $\mathbb{C}$. However, there are fields with positive characteristic (see Exercise 9.17), and the representation theory over such fields (so-called *modular representation theory*), even for uncomplicated groups, is still an active area of research. In this book, except for an occasional exercise, we only utilize fields of characteristic zero.

The most convenient field to use in most instances, and which we usually adopt, is the field of complex numbers $\mathbb{C}$, which is *algebraically closed*. That is, any polynomial equation with complex coefficients and positive degree has a solution in the complex numbers. In particular, this property guarantees the existence of eigenvalues of linear operators on a complex vector space, an attribute that is essential for some important results. Readers not fluent with complex numbers shouldn't panic; we will make only a few computations using them, and the moniker "complex," as well as "imaginary," is an unfortunate artifact of history.

Things start out pretty casually. In the first two chapters we develop the notions of a group, group actions, and group representations. Those who have completed a course in modern algebra should find this an easy read. Chapter 3 introduces some of the basic concepts and machinery of representation theory, and by the end of this chapter the reader should have a pretty good idea, supported by examples, of what "group representation theory" is about.

Chapter 4 examines the symmetric group in more detail. In Chap. 5, we give explicit decompositions of some polynomial spaces into irreducible representations, and Chap. 6 discusses two equivalent incarnations of the group algebra, a useful construction in representation theory and elsewhere.

The material now follows a more traditional exposition of the representation theory of the symmetric group, although we often relate it to representations on polynomial spaces. It also starts getting more technical, there is just no way to avoid it. You don't need to understand the proof of a theorem to apply the results, and novice readers can perhaps return later to understand the proofs and other technical details as their knowledge and abilities increase. Of course, it would be mathematical heresy to not include the proofs, and many of them are presented in a series of exercises—with solutions—that outline the basic ideas.

In Chap. 7 we discuss characters, a kind of function that "characterizes" a representation, and use these to label all of the irreducible representations of the symmetric group. Group characters are a useful tool in the study of group representations, and every book on representation theory should include a discussion

---

[1] The symbol := commonly denotes a definition.

of them, but characters are not essential to the more "constructive" approach that is our focus. In Chap. 8, we actually produce "realizations" of these representations in the group algebra and in polynomial spaces.

In Chap. 9 we introduce cosets, a basic construction in algebra, which are essential in order to construct representations of a group that are "induced" from that of a subgroup. Here, again, we present two (of several) equivalent versions of induced representations.

Chapter 10 describes direct products of groups, another standard algebraic construction, of which Young subgroups are an example. The trivial representations of Young subgroups are "induced" to representations called Young permutation modules. We also construct a new vector space of "polytabloids" that carries a representation of the symmetric group, and that has theoretical and computational advantages. We then relate these representations in polytabloid spaces to representations in polynomial spaces.

In Chap. 11, we construct the unique irreducible representation that appears in each Young permutation module, called a Specht module. Chapter 12 decomposes Young permutation modules into their irreducible components. Finally, in Chap. 13, we examine how the irreducible representations of S_n decompose when restricted to S_{n-1} or induced to S_{n+1}.

* An asterisk following an Exercise, Remark, etc., indicates that there is further information in the "Hints and Additional Comments" section at the end of each chapter. Of course, you should first try to work through them on your own.

Along the way, we will occasionally stray into related areas of mathematics that are suggested by our investigations. These are usually asterisked remarks placed at the end of the chapter so as not to break the logical flow of the main subject matter.

Contents

First Steps

1

We start by sketching out some of the basic concepts that are the subject of this book. We introduce the algebraic structure of a group, particularly permutation groups, as well as the concept of group actions, of which group representations are a special case.

1.1 Permutations and Groups

Most of us were first introduced to the concept of a permutation by something akin to the following elementary exercise:

> Billy, Joey, and Sally decide to start their very own club. How many different ways can they fill each of the offices of President, Vice-President, and Secretary?

After a little thought, the insightful reader realizes that at first there are three possible ways to choose the President, and once (say) Sally is so chosen, there remains only two possible ways to choose Vice President. Once Joey is chosen as VP then Billy is the only possible choice for Secretary, and thus there are $3 \times 2 \times 1 = 6$ such possible assignments.

Upon further reflection, it is easy to see that this question is equivalent to the question "How may ways can we arrange the list [Billy, Joey, Sally] in different orders?" Each such reordering of this list is called a *permutation* of the list, and in the realm of algebra, we are interested in the actual **action** of reordering. That is, we treat a permutation as a function.

Definition 1.1 Let A and B be sets.

- A function $f : A \to B$ is said to be *one-to-one*, or *injective*, or an *injection* if

$$f(a_1) = f(a_2) \text{ implies that } a_1 = a_2 \text{ for all } a_1, a_2 \in A.$$

© The Author(s), under exclusive license to Springer Nature Switzerland AG 2022
R. M. Howe, *An Invitation to Representation Theory*, SUMS Readings,
https://doi.org/10.1007/978-3-030-98025-2_1

- A function $f: A \to B$ is said to be *onto*, or *surjective*, or a *surjection* if

$$\text{for each } b \in B, \text{ we have } b = f(a) \text{ for some } a \in A.$$

- If $f: A \to B$ is both injective and surjective, then it is said to be *bijective*, or a *bijection*, or a *one-to-one correspondence*.

Definition 1.2 A *permutation* of a set S is a bijection from the set S to itself.

Although the set S need not be finite, we will restrict our consideration to finite S unless explicitly stated otherwise. Furthermore, we can capture the essential idea by assuming S to be the set $\{1, 2, 3, \ldots, n\}$ for some n.

Example 1.3 Let $S = \{1, 2, 3, 4\}$, and let γ be the permutation

$$\gamma(1) = 2, \ \gamma(2) = 4, \ \gamma(3) = 1, \ \gamma(4) = 3.$$

This is equivalent to rearranging the list $[1, 2, 3, 4]$ to $[2, 4, 1, 3]$.

Non-Example 1.4 The function f given by

$$f(1) = 2, \ f(2) = 2, \ f(3) = 1, f(4) = 4,$$

is not a permutation since it is not one-to-one.

Since permutations are functions from S to S, we can compose them in the usual way.

Example 1.5 Let $S = \{1, 2, 3, 4\}$, let γ be the permutation in Example 1.3 above, and let τ be the permutation

$$\tau(1) = 4, \ \tau(2) = 1, \ \tau(3) = 3, \ \tau(4) = 2.$$

Then

$$\tau \circ \gamma(1) = \tau(\gamma(1)) = \tau(2) = 1,$$
$$\tau \circ \gamma(2) = \tau(4) = 2,$$
$$\tau \circ \gamma(3) = \tau(1) = 4,$$
$$\tau \circ \gamma(4) = \tau(3) = 3.$$

Of course, we always have the trivial permutation (that doesn't change anything), and it should be apparent that, given any permutation σ of $\{1, 2, 3, \ldots, n\}$, we can find a permutation that "undoes" whatever σ does. If $\sigma(i) = j$, then the function that takes j back to i is such a permutation, denoted σ^{-1}. This observation is just the elementary fact that every bijection has an inverse that is also a bijection.

The above discussion can be mathematically summarized as follows; the collection of all permutations of a set, along with the operation of function composition, has the algebraic structure of a *group*.

Definition 1.6 A *group* is a set G with a binary operation (indicated by the symbol $*$ for now) that combines two elements from G to obtain another element in G, with the following properties:

(1) There is an element $e \in G$ (called the *identity* in G) such that

$$e * g = g * e = g \text{ for all } g \in G.$$

(2) For each $g \in G$, there is an element $g^{-1} \in G$ (called the *inverse* of g) such that

$$g^{-1} * g = g * g^{-1} = e.$$

(3) The operation $*$ is associative: $(g * h) * k = g * (h * k)$ for all $g, h, k \in G$.

Recall that a *binary operation* on a set is a way of combining two elements of the set to obtain another element of that set. This is often expressed by saying "The set is closed under the operation," and this criterion is customarily included in the definition of a group. When we need to be precise, we will denote a group G with operation $*$ as $(G, *)$, although it is typical to denote the group operation by juxtaposition, *i.e.*, $g * h$ is simply written as gh. The number of elements in a group G is called the *order* of G, written $|G|$, which need not be finite.

By the way, there are also unary operations (taking the transpose of a matrix), ternary operations (vector triple-product), and n-ary operations.

Example* 1.7 The integers with the operation of addition, $(\mathbb{Z}, +)$, is a group. Here the identity is 0 and the inverse of $k \in \mathbb{Z}$ is denoted $-k$. Since addition of integers is commutative, *i.e.*, $a + b = b + a$ for all $a, b \in \mathbb{Z}$, the group $(\mathbb{Z}, +)$ is said to be an *Abelian* group, named after the Norwegian mathematician Niels Abel.

Non-Example 1.8 The integers with the operation of multiplication. Certainly the integers are closed under multiplication, and we have the number 1 as the multiplicative identity, but not every integer has a multiplicative inverse **that is an integer**, and $0 \in \mathbb{Z}$ has no inverse.

Example 1.9 The set of all permutations of the set $\{1, 2, 3, \ldots, n\}$ with the operation of composition, and with inverses as described above. The identity element is the trivial permutation that does not change the order. This group is called the *symmetric group on n symbols*, and denoted S_n.

Example 1.10 One of the most important examples of a group is the set of invertible $n \times n$ matrices with the operation of matrix multiplication. This group is called the *general linear group of order n* and denoted $GL(n, \mathbb{F})$, where $\mathbb{F}$ is the field of scalars for a vector space (usually $\mathbb{R}$ or $\mathbb{C}$). This group is non-Abelian for $n \geq 2$.

Example 1.11 The set $\{-1, 1\}$ with the operation of multiplication.

Exercise 1.12 Work through and verify the assertions in the above examples.

Exercise* 1.13 We will see more efficient methods of notating permutations shortly. For now, try writing out some permutations and their inverses. Practice composing various permutations. If σ and ζ are arbitrary permutations, and if $\sigma \circ \zeta$ is their composition, what is $(\sigma \circ \zeta)^{-1}$? Does $\sigma \circ \zeta = \zeta \circ \sigma$? How many distinct permutations are there of a set with n elements?

Exercise 1.14 Find more examples (and non-examples!) of groups. Which of these groups are Abelian?

Exercise 1.15 Show that the identity element in a group is unique. That is, if there are two elements e and e' in G such that $g * e = e * g = g$, and $g * e' = e' * g = g$ for all $g \in G$, then $e = e'$. Thus the phrase "*the* identity in G" in Definition 1.6 is justified. Hint: What is $e * e'$?

Exercise* 1.16 Show that for each $g \in G$, the inverse g^{-1} is unique. That is, if there are two elements g^{-1} and g'^{-1} such that $gg^{-1} = g^{-1}g = e$ and $gg'^{-1} = g'^{-1}g = e$, then $g'^{-1} = g^{-1}$. Thus the phrase "*the* inverse of g" in Definition 1.6 is justified.

Exercise* 1.17 Let $(G, *)$ be a group, and let $g \in G$. For any positive integer k, define $g^k := \underbrace{g * g * \cdots * g}_{k \ times}$.

(1) Show that $g^k * g^l = g^{k+l}$.
(2) Show that $(g^k)^{-1} = (g^{-1})^k$, hence the notation g^{-k} is unambiguous.
(3) Verify some other familiar properties of exponents: $g^0 = e$, $[g^k]^l = g^{kl}$, etc.

1.2   Group Actions and Representations

There is a huge body of knowledge on the topic of "Group Theory." One of the important applications of groups is that they "act" on things.

Definition 1.18 We say that a group G *acts on a set S* (indicated by a lower dot) if:

(1) $g.s \in S$ for all $g \in G$ and $s \in S$.
(2) $g.(h.s) = (gh).s$ for all $g, h \in G$ and $s \in S$.
(3) $e.s = s$ for all $s \in S$, where $e \in G$ is the identity.

A set that is acted upon by a group G is sometimes referred to as a *G-set* or a *G-space*.

Example 1.19 The symmetric group S_n acts on the set $\{1, 2, 3, \ldots, n\}$ by $\sigma.i = \sigma(i)$ for $\sigma \in S_n$.

Example 1.20 Any group G acts on itself by left multiplication; $g.h := gh$ for $g, h \in G$.

Example 1.21 Any group G acts on itself by *conjugation*; $g.h := ghg^{-1}$ for $g, h \in G$. See Sect. 4.3.

Example 1.22 Let V be a vector space with ordered basis $\{e_1, e_2, e_3\}$. For example, $V = \mathbb{R}^3$ with standard basis

$$
e_1 = \begin{bmatrix} 1 \\ 0 \\ 0 \end{bmatrix}, \quad e_2 = \begin{bmatrix} 0 \\ 1 \\ 0 \end{bmatrix}, \quad \text{and} \quad e_3 = \begin{bmatrix} 0 \\ 0 \\ 1 \end{bmatrix}.
$$

Then S_3 acts on V by first permuting the basis vectors; $\sigma.e_i = e_{\sigma(i)}$ for $\sigma \in S_3$, which is then extended to all of V by linearity. That is, if $\sigma \in S_3$, if $r_1, r_2, r_3 \in \mathbb{R}$, and if $v = r_1 e_1 + r_2 e_2 + r_3 e_3 \in V$, then

$$
\sigma.v = \sigma.(r_1 e_1 + r_2 e_2 + r_3 e_3) := r_1 \sigma.e_1 + r_2 \sigma.e_2 + r_3 \sigma.e_3 = r_1 e_{\sigma(1)} + r_2 e_{\sigma(2)} + r_3 e_{\sigma(3)}.
$$

In this case we say that S_3 *acts linearly* on V.

Definition 1.23 If a group G acts linearly on a vector space V, we say that V is a *representation* of G. A mathematically more precise definition of representation is given in Sect. 1.5.

Example 1.24 The group $GL(n, \mathbb{R})$ acts on the vector space $\mathbb{R}^n$ by left matrix multiplication of column vectors (or by right matrix multiplication of row vectors). This is often referred to as the *defining representation* of $GL(n, \mathbb{R})$.

Remark 1.25 The representation in Example 1.22, where S_n permutes the basis vectors of an n-dimensional vector space, is called a *permutation representation* of S_n.

Example 1.26 Let $\sigma \in S_3$, where $\sigma(1) = 2, \sigma(2) = 3$, and $\sigma(3) = 1$. Let V be a 3-dimensional vector space with ordered basis $\{\mathbf{e}_1, \mathbf{e}_2, \mathbf{e}_3\}$. Let $\mathbf{v} = r_1\mathbf{e}_1 + r_2\mathbf{e}_2 + r_3\mathbf{e}_3$, and let $(r_1, r_2, r_3) \in \mathbb{R}^3$ be the coordinate vector of $\mathbf{v}$. Then $\sigma.\mathbf{v} = r_1\mathbf{e}_2 + r_2\mathbf{e}_3 + r_3\mathbf{e}_1$ has coordinate vector (r_3, r_1, r_2). Thus S_3 acts of the vector space of coordinate-vectors as $\sigma.(r_1, r_2, r_3) = (r_3, r_1, r_2) = (r_{\sigma^{-1}(1)}, r_{\sigma^{-1}(2)}, r_{\sigma^{-1}(3)})$.

Definition 1.27 If a group G acts on a set S, and if $s \in S$, then the set $\{g.s \mid g \in G\}$ is called the *orbit* of s, denoted $o(s)$. We say the action of G on S is *trivial* if $o(s) = \{s\}$ for all $s \in S$. That is, if $g.s = s$ for all $g \in G$ and all $s \in S$. We say that the action of G on S is *transitive* if $o(s) = S$ for all $s \in S$. That is, for every s and s' in S there is a $g \in G$ such that $g.s = s'$.

Example 1.28 The action of S_n on the set $\{1, 2, \ldots, n\}$ is transitive, as is the action in Example 1.20. The action of $GL(n, \mathbb{R})$ is transitive on the set of non-zero vectors in $\mathbb{R}^n$. The action of a group G on itself by conjugation in Example 1.21 is never transitive (unless G consists of only the identity element), and is trivial if G is Abelian.

Exercise* 1.29 If a group G acts on a set S, show that the G-orbits in S are equivalence classes.

Remark* 1.30 Equivalence classes arise often in mathematics, and we will see several examples later in the form of quotient spaces, cosets, etc. It is important when working with equivalence classes that things be well-defined.

1.3   More About the Symmetric Group

There are several ways to describe permutations and compute their compositions. One way is to write a permutation, such as γ from Example 1.3, as an array with $\gamma(i)$ placed below each i. The order of the entries in the top row of the array doesn't matter. For this example we have,

$$\gamma = \begin{bmatrix} 1 & 2 & 3 & 4 \\ 2 & 4 & 1 & 3 \end{bmatrix}.$$

Similarly, we can write τ from Example 1.5 as

$$\tau = \begin{bmatrix} 1 & 2 & 3 & 4 \\ 4 & 1 & 3 & 2 \end{bmatrix}.$$

We can then compose the two permutations by placing one array over the other;

$$\tau\gamma = \begin{bmatrix} 1\ 2\ 3\ 4 \\ 2\ 4\ 1\ 3 \end{bmatrix}$$
$$\begin{bmatrix} 2\ 4\ 1\ 3 \\ 1\ 2\ 4\ 3 \end{bmatrix}$$

$$= \begin{bmatrix} 1\ 2\ 3\ 4 \\ 1\ 2\ 4\ 3 \end{bmatrix}.$$

A more concise and fruitful way to designate a permutation is to use *cycle notation*, which is also useful theoretically. If $\sigma \in S_n$, and if $k \in \{1, 2, \ldots, n\}$, then a *cycle* is a list

Equation 1.31

$$(k, \sigma(k), \sigma^2(k), \ldots, \sigma^m(k), \sigma^{m+1}(k) = k).$$

Here of course $\sigma^2(k) = \sigma \circ \sigma(k) = \sigma(\sigma(k))$ and so forth. The entry $\sigma^{m+1}(k) = k$ is understood and so it is omitted, and the elements $k, \sigma(k), \sigma^2(k), \ldots, \sigma^m(k)$ are all distinct. For example, the permutation

$$\sigma = \begin{bmatrix} 1\ 2\ 3\ 4\ 5\ 6\ 7 \\ 2\ 3\ 4\ 1\ 6\ 7\ 5 \end{bmatrix}$$

contains two cycles; one cycle of length four, *i.e*, a "four-cycle," $(1, 2, 3, 4)$, and one three-cycle, $(5, 6, 7)$. We then write $\sigma = (1, 2, 3, 4)(5, 6, 7)$, which means that "$\sigma$ takes one to two, two to three, three to four, and four to one; σ takes five to six, six to seven, and seven to five."

Similarly, from Example 1.5 we have $\gamma = (1, 2, 4, 3)$ and $\tau = (1, 4, 2)(3)$. If there is no confusion, it is customary to omit writing down the one-cycles, such as (3) in τ above, and the omitted one-cycles are understood to indicate the elements of $\{1, 2, \ldots, n\}$ that are fixed by the permutation. Of course we can't omit everything, so the identity permutation (which fixes everything) is written as (1) in cycle notation. It is also usual to write the smallest number first, for example $(2, 4, 1, 3) = (1, 3, 2, 4)$. Note that $\tau\gamma = (1, 4, 2)(1, 2, 4, 3) = (3, 4)$, and that the operation is performed right-to-left, as is typical of function composition.

Example 1.32 In cycle notation, the elements of S_3 are

$$S_3 = \{(1), (1, 2), (1, 3), (2, 3), (1, 2, 3), (1, 3, 2)\}.$$

Notation 1.33 We will use the notation e for the identity when we are talking about groups in general, or e_G if we need to indicate the identity for a specific group G. The identity element e_{S_n} will usually be written as (1) using cycle notation, and we trust the reader can distinguish from the context when we are talking about the identity permutation (1), versus (1), the first item in an enumerated list.

Exercise 1.34 For each $\sigma \in S_3$, determine σ^{-1} using cycle notation.

Exercise 1.35 Write down all the elements in the groups S_2 and S_4 using cycle notation. Determine some inverses, and practice calculating some compositions.

The following exercises are covered in detail in Chap. 4, but you should give them a try here.

Exercise* 1.36 Prove the assertion implicit in Eq. 1.31: for each $\sigma \in S_n$, and each $k \in \{1, 2, \ldots n\}$, there is some positive integer m, with $0 \leq m \leq n - 1$, such that $\sigma^{m+1}(k) = k$.

Exercise* 1.37 Two cycles are said to be *disjoint* if they have no entries in common. Prove that any permutation can be written as a product of disjoint cycles. Prove that disjoint cycles commute.

1.4   More Groups and Subgroups

A common theme in mathematics is that various mathematical structures have "substructures." Familiar examples include subsets of a set and subspaces of a vector space, but the list goes on: subspaces of a topological space, submanifolds of a manifold, etc. Here we introduce subgroups of a group, and shortly we will see subrepresentations of a group representation.

Example 1.38 Check that the set of even integers with addition, denoted $(2\mathbb{Z}, +)$, is a group. Note that $2\mathbb{Z} \subset \mathbb{Z}$.

Definition 1.39 A subset H of a group G that is itself a group (with the same operation as G) is called a *subgroup* of G, often denoted $H < G$.

Non-Example* 1.40 The set of odd integers under addition is not a subgroup of $(\mathbb{Z}, +)$.

Non-Example 1.41 The set of positive real numbers under multiplication, denoted $(\mathbb{R}^+, \times)$, is a group, but is not a subgroup of $(\mathbb{R}, +)$ since the operations are different.

Example 1.42 If G is a group, and if e is the identity element in G, then $\{e\}$ is a subgroup of G, called the *trivial subgroup*.

Exercise* 1.43 Generalize Example 1.38. For some fixed integer k, let $k\,\mathbb{Z}$ denote all integer multiples of k. Check that $(k\,\mathbb{Z}, +)$ is a subgroup of $(\mathbb{Z}, +)$. If k_1 and k_2 are integers, when is $k_1\,\mathbb{Z}$ a subgroup of $k_2\,\mathbb{Z}$? For example, $3\,\mathbb{Z}$ is not a subgroup of $2\,\mathbb{Z}$, but $6\,\mathbb{Z}$ is a subgroup of both $2\,\mathbb{Z}$ and $3\,\mathbb{Z}$.

Exercise* 1.44 Let $(\mathbb{R}^2, +)$ be the set of ordered pairs of real numbers with addition defined component-wise: $(a, b) + (c, d) = (a+c, b+d)$. Verify that the set $\{(x, 2x) \mid x \in \mathbb{R}\}$ is a subgroup of $(\mathbb{R}^2, +)$. Verify that the set $\{(x, 2x+1) \mid x \in \mathbb{R}\}$ is NOT a subgroup. Generalize these results.

Exercise 1.45 Show that the set of all functions of the form $\{f(x) = ax + b \mid a, b \in \mathbb{R}\}$ is a group, and that the set $\{f(x) = ax \mid a \in \mathbb{R}\}$ is a subgroup.

Exercise 1.46 Check that the set $\{1, i, -1, -i\}$ is a subgroup of the group $(\mathbb{C}^*, \times)$, the non-zero complex numbers with multiplication. Observe that the group from Example 1.11 is a subgroup of this subgroup. In case there's any confusion here, i is the "imaginary unit": $i^2 = -1$.

Exercise 1.47 Show that the set $\{a + bi \mid a, b \in \mathbb{R}, \ a^2 + b^2 = 1\}$ is a subgroup of $(\mathbb{C}^*, \times)$, the non-zero group of complex numbers under multiplication, with $\{1, i, -1, -i\}$ as a subgroup.

There is an easy way to check for subgroups using the "One-Step Subgroup Test."

Theorem 1.48 *Let G be a group, and let H be a nonempty subset of G. If $a, b \in H$ implies that $ab^{-1} \in H$, then H is a subgroup of G.*

Exercise* 1.49 Prove Theorem 1.48.

Exercise* 1.50 Let g be an element of a group G. Show that the set

$$\langle g \rangle := \{\dots, g^{-2}, g^{-1}, e, g, g^2, \dots\} = \{g^k \mid k \in \mathbb{Z}\}$$

is an Abelian subgroup of G, called the *cyclic subgroup generated by g*. Show that if $|G|$ is finite, then $g^k = e$ for some k. For example, the group in Exercise 1.46 is a finite cyclic group generated by i. Can you find other examples?

Most of us have an intuitive notion of the meaning of the term "symmetry" as "when two or more parts of an object or figure are identical after a flip or turn." These next examples demonstrate the mathematical notion of a group makes precise this informal notion of symmetry.

Example 1.51 Label the vertices of an equilateral triangle clockwise with the numbers $1, 2, 3$. Show that the symmetries of the triangle form a group, and that the elements in the group can be labeled by permutations. For example, counterclockwise rotation by $60°$ corresponds to the permutation $(1, 2, 3)$, and reflection across the line from vertex number 1 to the opposite side corresponds to the permutation $(2, 3)$. If you write out all the elements using cycle notation, this group should look familiar.

Example 1.52 Label the vertices of a square clockwise with the numbers $1, 2, 3, 4$. Show that the symmetries of the square can be labeled by permutations, and form a group that is a subgroup of S_4. For example, counterclockwise rotation by $90°$ corresponds to the permutation $(1, 2, 3, 4)$, and reflection across the line from vertex 1 to vertex 3 corresponds to the permutation $(2, 4)$. This group is called D_4, the *dihedral group.*

Exercise* 1.53 Write out the elements of D_4 in cycle notation. What are the subgroups of D_4, and which subgroups are subgroups of other subgroups? Hint: $|D_4| = 8$.

Exercise* 1.54 Show that if (i, j) is any two-cycle in S_n, then $\{(1), (i, j)\}$ is a subgroup of S_n.

Exercise* 1.55 Write out all of the subgroups of S_3, S_4 and S_5.

Exercise* 1.56 Let $SL(n, \mathbb{R})$ denote the set of real $n \times n$ matrices with determinant equal to one (*the special linear group*). Show that $SL(n, \mathbb{R})$ is a subgroup of $GL(n, \mathbb{R})$. Hint: Use the properties of the determinant function and Theorem 1.48.

Exercise* 1.57 Verify that $O(n, \mathbb{R})$, the set of $n \times n$ orthogonal matrices, is a subgroup of $GL(n, \mathbb{R})$ called the *orthogonal group*. Recall that a matrix A is orthogonal if $A^T A = A A^T = I$, where A^T denotes the matrix transpose of A, and where I is the identity matrix. Let $O(n, \mathbb{R})$ act on $\mathbb{R}^n$ by left multiplication of column vectors in the usual way. What is the $O(n, \mathbb{R})$-orbit of a vector in $\mathbb{R}^n$?

Exercise 1.58 Show that if $A \in O(n, \mathbb{R})$, then $\text{Det } A = \pm 1$.

Exercise 1.59 This is a standard example from linear algebra. Let R_θ be the rotation of a vector

$$\mathbf{v} = \begin{bmatrix} x \\ y \end{bmatrix} \in \mathbb{R}^2$$

counterclockwise by an angle θ. Make a geometric argument to show that $R_\theta \colon \mathbb{R}^2 \to \mathbb{R}^2$ is linear. Show that the action of R_θ on $\mathbf{v}$ is given by the matrix product

$$R_\theta\,(\mathbf{v}) = \begin{bmatrix} \cos\theta & -\sin\theta \\ \sin\theta & \cos\theta \end{bmatrix} \begin{bmatrix} x \\ y \end{bmatrix}.$$

Exercise 1.60 Let $SO(n, \mathbb{R})$ denote the *special orthogonal group*. That is,

$$SO(n, \mathbb{R}) := \{A \in O(n, \mathbb{R}) \mid \mathrm{Det}\, A = 1\} = O(n, \mathbb{R}) \cap SL(n, \mathbb{R}).$$

Show that $SO(n, \mathbb{R})$ is a subgroup of $O(n, \mathbb{R})$ (or of $SL(n, \mathbb{R})$ or $GL(n, \mathbb{R})$ for that matter). Verify that the matrix R_θ in Exercise 1.59 is an element of $SO(2, \mathbb{R})$, and conclude that $SO(2, \mathbb{R})$ is the group of rotations in the plane.

Exercise 1.61 Let A be an $n \times n$ matrix with complex entries, and let $A^\dagger$ denote the conjugate transpose of A. A is said to be *unitary* if $A^\dagger A = AA^\dagger = I$. Show that $U(n, \mathbb{C})$, the set of all $n \times n$ unitary matrices, is a subgroup of $GL(n, \mathbb{C})$, called the *unitary group*.

Exercise 1.62 If a group G acts on a set S, and if $s \in S$, then the set

$$G_s := \{g \in G \mid g.s = s\}$$

is called the *stabilizer* of s, or the *stability subgroup* of s. Show that, in fact, G_s is a subgroup of G.

Exercise 1.63 Let G be a group, and let $\mathcal{Z}G$ denote the *center* of G;

$$\mathcal{Z}G := \{h \in G \mid hg = gh \text{ for all } g \in G\}.$$

In other words, $\mathcal{Z}G$ consists of those elements in G that commute with every other element in G. Show that $\mathcal{Z}G$ is a subgroup of G.

Exercise* 1.64 Show that $\mathcal{Z}GL(n, \mathbb{C})$ consists of the scalar matrices, that is, scalar multiples of the identity matrix.

1.5 Group Homomorphisms and More About Representations

Another common theme in mathematics is that we have certain types of functions, or "mappings," that preserve a given mathematical structure. For example, in linear algebra we have the notion of a linear transformation. For vector spaces V and W, a map $T \colon V \to W$ is *linear* if $T(r\mathbf{v} + s\mathbf{u}) = rT(\mathbf{v}) + sT(\mathbf{u})$ for all $\mathbf{v}, \mathbf{u} \in$

V and all scalars r, s. A linear transformation that is also a bijection (one-to-one and onto, and therefore invertible) is said to be an *isomorphism*. An isomorphism from a vector space V to itself is said to be an *automorphism*, and the set of all such automorphisms of V with the operation of composition is easily seen to be a group, denoted $GL(V)$ (the general linear group on V), or sometimes $Aut(V)$. Two vector spaces are said to be *isomorphic* if there is an isomorphism between them. Isomorphic vectors spaces are essentially the same, and it is a standard result that any two finite-dimensional vector spaces are isomorphic exactly when they have the same dimension.

Linear maps preserve the vector space structure, and consequently yield information about vector spaces in general. The analogue of this concept for groups are called *group homomorphisms*. There are homomorphisms of other algebraic structures (rings, modules, etc.), but we will drop the "group" designation when there is no ambiguity.

Definition 1.65 Let G and H be groups. A map $\phi : G \to H$ is a *group homomorphism* if $\phi(ab) = \phi(a)\phi(b)$ for all $a, b \in G$. If ϕ is also a bijection, then ϕ is said to be a *group isomorphism*. An isomorphism from a group G to itself is called an *automorphism* of G.

Example 1.66 The determinant function Det: $GL(n, \mathbb{R}) \to (\mathbb{R}^*, \times)$ is a homomorphism. Here $(\mathbb{R}^*, \times)$ is the group of non-zero real numbers with multiplication.

Example 1.67 For any integer k, the map $\phi_k : (\mathbb{Z}, +) \to (\mathbb{Z}, +)$ given by $\phi_k(n) = kn$ is a homomorphism. Note that, in the case of additive notation for the group operation, we require that $\phi_k(n + m) = \phi_k(n) + \phi_k(m)$.

Example 1.68 Let $(\mathbb{R}, +)$ denote the set of real numbers under addition. For any real number t, the map $\phi_t : (\mathbb{R}, +) \to (\mathbb{R}, +)$ given by $\phi_t(r) = tr$ is a homomorphism.

Non-Example 1.69 For any integer $k \neq 0$, the map $\psi_k : (\mathbb{Z}, +) \to (\mathbb{Z}, +)$ given by $\psi_k(n) = k + n$ is a not a homomorphism.

Example 1.70 Let e denote the natural exponential base familiar from calculus (not the group identity element). The map that takes $(\mathbb{R}, +) \to (\mathbb{R}^+, \times)$ given by $t \mapsto e^t$ is a homomorphism, where $(\mathbb{R}^+, \times)$ is the group of positive real numbers with multiplication. This generalizes; if r is a positive real number, then the map $t \mapsto r^t$ is a homomorphism.

Example 1.71 For those familiar with complex variables, the map

$$t \mapsto e^{it} = \cos t + i \sin t$$

is a homomorphism from the group $(\mathbb{R}, +)$ to the group from Exercise 1.47.

Example 1.72 The map that takes $(\mathbb{R}, +) \to SL(2, \mathbb{R})$ (Exercise 1.56) given by $t \mapsto \left(\begin{smallmatrix} 1 & t \\ 0 & 1 \end{smallmatrix} \right)$ is a homomorphism.

Example 1.73 Given a finite dimensional $\mathbb{R}$-vector space V with an ordered basis, the two groups $GL(n, \mathbb{R})$ and $GL(V)$ are isomorphic.

Exercise* 1.74 Verify the above examples.

Exercise* 1.75 Consider the subset of elements in S_n that stabilize n. Show that this is a subgroup of S_n that is isomorphic to S_{n-1}. This is usually referred to as the *standard embedding* of S_{n-1} in S_n. Check that the subset of elements in S_n that fix one of the numbers $1, 2, \ldots, n$ is a subgroup of S_n that is also isomorphic to S_{n-1}.

Remark 1.76 Unless explicitly stated otherwise, whenever we discuss S_k as a subgroup of S_n for $k < n$, we mean the the subgroup of S_n that permutes only the elements $\{1, 2, \ldots, k\}$ and fixes the elements $\{k + 1, \ldots, n\}$.

Remark 1.77 Example 1.20 says that the action of left multiplication by a group G on itself permutes the elements of G. Thus any finite group is isomorphic to a subgroup of S_n for some n. This result is known as *Cayley's theorem*.

Exercise* 1.78 Show that S_n in non-Abelian for $n \geq 3$.

Here are two important subgroups related to homomorphisms.

Definition 1.79 If G and H are groups, and if $\phi: G \to H$ is a homomorphism, define the *kernel* of ϕ as Ker $\phi := \{g \in G \mid \phi(g) = e_H\}$, where e_H is the identity element in H. Define the *image* of ϕ to be Im $\phi := \{h \in H \mid h = \phi(g)$ for some $g \in G\}$.

Example 1.80 The subgroup $SL(n, \mathbb{R})$ of $GL(n, \mathbb{R})$ is the kernel of the Determinant map from Example 1.66.

Example 1.81 The homomorphism from Example 1.71 has kernel $\{2k\pi \mid k \in \mathbb{Z}\}$.

Example 1.82 The map from $(\mathbb{Z}, +)$ to the group $(\{1, i, -1, -i\}, \times)$ (Exercise 1.46) given by $k \mapsto i^k$ is a homomorphism with kernel $(4\mathbb{Z}, +)$.

Example 1.83 Let $(\mathcal{P}(x), +)$ denote the group of polynomial functions (in one variable) with real coefficients, and with the operation of addition. Since $(f + g)' = f' + g'$, the derivative is a group homomorphism from $(\mathcal{P}(x), +)$ to $(\mathcal{P}(x), +)$. Since the zero-polynomial is the identity element, the kernel of this map is the set of constant polynomials.

Exercise 1.84 If $\phi : G \to H$ is a group homomorphism, verify that Ker ϕ and Im ϕ are subgroups of G and H respectively. Show that ϕ is injective if and only if Ker $\phi = \{e\}$. Show that $\phi(g^{-1}) = [\phi(g)]^{-1}$. Hint: Recall the proof of the analogous propositions for linear maps between vector spaces. You should also work through and verify the above examples.

We can now provide a more mathematically precise definition of a representation, as promised in Definition 1.23.

Definition 1.85 A *representation* of a group G on a vector space V is a homomorphism $\rho : G \to GL(V)$. The dimension of V is called the *degree* of the representation. If $\rho(g)\mathbf{v} = \mathbf{v}$ for all $g \in G$ and for all $\mathbf{v} \in V$, then the representation is said to be *trivial*. If ρ is injective (one-to-one), the representation is said to be *faithful*.

We emphasize that when an element $g \in G$ acts linearly on a vector space V, then $\rho(g)$ is an invertible linear map from V to itself. When we need to be precise, we will write (ρ, V) for the representation $\rho(g)\mathbf{v}$ for $g \in G$ and $\mathbf{v} \in V$. Where there is no chance for ambiguity, it is customary to use the "lower dot" notation and write $g.\mathbf{v}$ for the action of g on $\mathbf{v}$. It is common, if somewhat imprecise, to just say "V is a representation of G." The terminology "V is a G-module" or "V is a G-space" is also seen in the literature, and sometimes we see the expression "V carries a representation of G."

Exercise* 1.86 Let $V = \mathbb{R}^3$. Consider the permutation representation, where S_3 acts on V by permuting the standard basis (column) vectors as in Example 1.22. Write out the matrices for this group action. Hint: if $GL(n, \mathbb{R})$ acts on $\mathbb{R}^n$ by left multiplication of column vectors, and if σ is a permutation in S_3 with $\widetilde{\sigma}$ the corresponding matrix, then the ith column of $\widetilde{\sigma}$ is the image of the ith basis vector. The resulting matrices should consist of only zeros and ones, with a unique one in each row and column. Such matrices are referred to as *permutation matrices*.

Definition* 1.87 Once bases are chosen, any linear map between any two finite-dimensional vector spaces can be considered as a matrix acting by multiplication on the coordinate vectors, and we call this the *matrix realization* of the linear map. Exercise 1.59 gives the matrix realization of the rotation R_θ. Similarly, the matrices such as $\widetilde{\sigma}$ from Exercise 1.86 are called *matrix realizations* of the representation. When we need to be more precise we will use the notation $\widetilde{\rho}(\sigma)$.

Exercise* 1.88 Let $G = (\mathbb{R}^*, \times)$ be the group of non-zero real numbers under multiplication, and let $\mathbb{P}_2$ denote the vector space of polynomials (in one variable) of degree two or less. Verify that the action of G on $\mathbb{P}_2$ given by $a.p(x) = p(a^{-1}x)$ is a representation of G on $\mathbb{P}_2$. What is the matrix realization of this representation with respect to the standard basis $\{1, x, x^2\}$ for $\mathbb{P}_2$?

Exercise 1.89 If a real vector space V is a representation of G, revue from linear algebra that the map $g \mapsto \tilde{g}$ is a group homomorphism between G and $GL(n, \mathbb{R})$.

Exercise* 1.90 Recall from linear algebra that rotations and reflections in $\mathbb{R}^2$ are linear maps from $\mathbb{R}^2$ to $\mathbb{R}^2$ (see Exercise 1.59). Write out the matrix realization of the symmetries of the equilateral triangle from Exercise 1.51 with respect to the standard basis for $\mathbb{R}^2$. You might also try this for the symmetries of the square in Exercise 1.52. Verify that these matrices are all unitary (Exercise 1.61), which is the same as orthogonal since the entries are real numbers.

Exercise 1.91 Let G be the group $\{1, i, -1, -i\}$ from Exercise 1.46. Verify that the map $\phi : G \to GL(2, \mathbb{R})$ given by

$$\phi(1) = \begin{pmatrix} 1 & 0 \\ 0 & 1 \end{pmatrix}, \quad \phi(i) = \begin{pmatrix} 0 & -1 \\ 1 & 0 \end{pmatrix}, \quad \phi(-1) = \begin{pmatrix} -1 & 0 \\ 0 & -1 \end{pmatrix}, \quad \text{and} \quad \phi(-i) = \begin{pmatrix} 0 & 1 \\ -1 & 0 \end{pmatrix}$$

is a group homomorphism, and therefore ϕ is a representation of G on $\mathbb{R}^2$.

1.6 Representations on Function Spaces

If S is a set, and if $\mathbb{F}$ is a field of scalars, then the collection of scalar-valued functions on S,

$$\mathbb{F}[S] := \{f \mid f : S \to \mathbb{F}\},$$

is a vector space. Addition and scalar multiplication in $\mathbb{F}[S]$ is defined by

$$[f + g](x) := f(x) + g(x), \quad \text{and} \quad [rf](x) := r[f(x)], \quad r \in \mathbb{F}.$$

In other words, addition of functions is defined via addition of scalars in the target-space, and similarly for scalar multiplication. If G acts on S, then we can define a representation of G on the vector space $\mathbb{F}[S]$ by

$$[g.f](x) := f(g^{-1}.x) \text{ for } f \in \mathbb{F}[S], \ g \in G, \ \text{and} \ x \in S.$$

This is sometimes referred to as "linearization of a symmetry," or "linearization of a group action."

Exercise 1.92 Verify that this action on a function space is, in fact, a representation. Hint: Note that $[g.[h.f]](x) = [h.f](g^{-1}x)$.

Remark 1.93 If V is a finite-dimensional vector space, then the collection of scalar-valued linear functions on V is referred to as the *dual space* of V, usually[1] denoted V^*. It is a standard linear algebra exercise to show that if $\{e_i\}_{i=1}^n$ is a basis for V, and if we define $e_i^* \in V^*$ by[2]

$$e_i^*(e_j) = \delta_{ij} := \begin{cases} 1 & \text{if } i = j; \\ 0 & \text{otherwise,} \end{cases}$$

then the set $\{e_i^*\}_{i=1}^n$ is a basis for V^*. This basis is called the *dual basis* to $\{e_i\}_{i=1}^n$.

Exercise* 1.94 Verify that the dual basis described in Remark 1.93 is, in fact, a basis for V^*. Conclude that $\text{Dim } V = \text{Dim } V^*$, and therefore V and V^* are isomorphic as vector spaces.

Exercise* 1.95 If $\{e_i\}_{i=1}^n$ is an ordered basis for a vector space V, and if

$$\mathbf{v} = r_1 e_1 + r_2 e_2 + \cdots + r_n e_n \in V,$$

then the ith coordinate function x_i is defined by $x_i(\mathbf{v}) = r_i$. Verify that these coordinate functions are the dual basis to the given basis on V.

Example 1.96 (Very Important) If $\sigma \in S_n$ acts on a vector space V by permuting the basis vectors, and if x_i is the ith coordinate function on V, then

$$[\sigma.x_i](\mathbf{v}) = x_i(\sigma^{-1}.\mathbf{v}) = x_{\sigma(i)}(\mathbf{v}) \text{ for all } \mathbf{v} \in V.$$

To be clear, the notation $[\sigma.x_i]$ means the representation of S_n on the coordinate functions x_i, while the notation $\sigma^{-1}.\mathbf{v}$ denotes the permutation representation of S_n on V.

---

[1] If $\mathbb{F}$ is the field of scalars for V, then $\text{Hom}_{\mathbb{F}}(V, \mathbb{F})$ is also common notation for the dual space.

[2] The symbol δ_{ij}, a function of the two variables i and j, is called the *Kronecker delta*.

More explicitly, if $\sigma \in S_n$ and $\mathbf{e}_j$ is any basis vector in V, then

$$[\sigma.x_i](\mathbf{e}_j) = x_i(\sigma^{-1}.\mathbf{e}_j)$$

$$= x_i(\mathbf{e}_{\sigma^{-1}(j)})$$

$$= \begin{cases} 1 \text{ if } i = \sigma^{-1}j \\ 0 \text{ otherwise} \end{cases}$$

$$= \begin{cases} 1 \text{ if } \sigma(i) = j \\ 0 \text{ otherwise} \end{cases}$$

$$= x_{\sigma(i)}(\mathbf{e}_j).$$

Thus $\sigma.x_i = x_{\sigma(i)}$.

Remark* 1.97 There is a related action of S_n, called a *place-permutation* action, in which S_n acts on positions or places.

1.7   Hints and Additional Comments

Example 1.7 Recall that an operation $*$ on a set S is *commutative* if $a * b = b * a$ for all $a, b \in S$. For historical reasons, when discussing groups we use the term *Abelian*, after Niels Abel. As such, the term "Abelian" should be capitalized in the literature, but often is not.

Exercise 1.13 Check that the two permutations γ and τ from Example 1.5 do not commute.

If σ and ζ are arbitrary permutations, then $(\sigma \circ \zeta) \circ (\zeta^{-1} \circ \sigma^{-1}) = \sigma \circ (\zeta \circ \zeta^{-1}) \circ \sigma^{-1} = e$. Thus

$$(\sigma \circ \zeta)^{-1} = \zeta^{-1} \circ \sigma^{-1},$$

as is the case for all non-commutative operations, such as multiplication of invertible matrices. Note that associativity is required.

If there are n elements in a set, then there are n ways to choose the first element, leaving $n - 1$ ways to choose the second element, $n - 2$ ways to choose the third, etc. Thus there are $n!$ distinct permutations on a set with n elements. In other words, $|S_n| = n!$.

Exercise 1.16 As in Exercise 1.15, suppose that there is some other element, say g'^{-1}, such that $gg'^{-1} = g'^{-1}g = e$. Then

$$\begin{aligned}
g^{-1} &= g^{-1}e \\
&= g^{-1}(gg'^{-1}) \\
&= (g^{-1}g)g'^{-1} \\
&= eg'^{-1} \\
&= g'^{-1}.
\end{aligned}$$

Note that associativity is required. In non-associative settings there can be distinct left and right inverses, as well as distinct left and right identities.

Exercise 1.17 This is virtually identical to the case for invertible matrices, which can be found in any linear algebra book. The most complete proof would use induction on k.

Exercise 1.29 A *relation* on a set is a way of pairing (relating) two elements from the set. For example, the inequality $a \leq b$ for $a, b \in \mathbb{R}$ is a way of relating two real numbers. An *equivalence relation*, typically denoted $\sim$, is a relation that generalizes the notion of equality. A relation $\sim$ is an equivalence relation on a set S if

- $a \sim a$ for all $a \in S$ (reflexive);
- $a \sim b \Rightarrow b \sim a$ (symmetric);
- $a \sim b$ and $b \sim c \Rightarrow a \sim c$ (transitive).

The relation $\leq$ on $\mathbb{R}$ is reflexive and transitive, but not symmetric. One example of an equivalence relation that you should be familiar with from linear algebra is the relation of similar $n \times n$ matrices.

The *equivalence class* of $a \in S$, denoted $cl(a)$, is the set of all elements in S that are equivalent to a. That is, $cl(a) := \{b \in S \mid b \sim a\}$. For the case at hand, we wish to show that "is in the same G-orbit" is an equivalence relation, and consequently $o(a) = cl(a)$

Now "a is in the same G-orbit as b" means that $a = g.b$ for some $g \in G$. But then $g^{-1}.a = b$, so "b is in the same orbit as a," and the relation is reflexive. If $a = g.b$, and if $c = h.b$, then $a = gh^{-1}.c$, and thus a is in the same orbit as c, so the relation is transitive. Obviously a is in the same orbit as a since $a = e.a$, so the relation is reflexive.

Remark 1.30 An elementary example of an equivalence relation is that of equality of two rational numbers in $\mathbb{Q}$. If a, b, c, d are integers (with $b, d \neq 0$), then we say that

$$\frac{a}{b} = \frac{c}{d} \text{ if and only if } ad = bc, \text{ and thus } \frac{1}{2} = \frac{3}{6}, \text{ etc.}$$

If we were to define a "function" $f : \mathbb{Q} \to \mathbb{Z}$ by $f(\frac{a}{b}) = a$, then $f(\frac{1}{2}) = 1$, but $f(\frac{3}{6}) = 3$, so this "function" is not well-defined on equivalence classes. Similarly, the "operation"

$$\frac{a}{b} + \frac{c}{d} = \frac{a+c}{b+d}$$

is not well-defined on equivalence classes (check this), much to the annoyance of 9 year old students everywhere.

Exercise 1.36 The relevant observation here is that the list

$$k, \sigma(k), \sigma^2(k), \ldots, \sigma^m(k), \ldots$$

can contain only finitely many distinct elements. For details see Proposition 4.2.

Exercise 1.37 Apply Exercise 1.36, then apply it again to any element not in the first cycle, etc. Since the cycles are disjoint, no letter (on which the permutations act) can be affected more than once. For more details see Proposition 4.1.

Non-Example 1.40 Observe, for example, that $3 + 5 = 8$, so the set of odd integers is not closed under addition.

Exercise 1.43 Verifying that $k\mathbb{Z}$ is a subgroup of $\mathbb{Z}$ is a routine application of the definitions involved. The set $k_1\mathbb{Z}$ is a subgroup of $k_2\mathbb{Z}$ if and only if k_2 divides k_1.

Exercise 1.44 The subset $\{(x, ax + b) \mid x \in \mathbb{R}\}$, (for a and b fixed elements of $\mathbb{R}$) is a subgroup of $(\mathbb{R}^2, +)$ if and only if $b = 0$. One easy way to see this is that the additive identity $(0, 0)$ must be an element of any subgroup.

Exercise 1.49 Since H is non-empty, there is some $x \in H$, and by hypothesis, $xx^{-1} = e \in H$, so H contains the identity. Since $e \in H$ and $x \in H$, by hypothesis so is $ex^{-1} = x^{-1} \in H$, so for any $x \in H$ there is an $x^{-1} \in H$. If there is another element $y \in H$, we have shown that $y^{-1} \in H$ and by hypothesis $xy = x(y^{-1})^{-1} \in H$, so H is closed under the operation, which must be associative since it is the same operation as G.

Exercise 1.50 As in Exercise 1.36, if G is finite, then the elements in

$$\{\ldots, g^{-2}, g^{-1}, e, g, g^2, \ldots\}$$

cannot all be distinct, so $g^m = g^n$ for some $n > m$, and thus $g^k = g^{n-m} = e$. The smallest such positive k such that $g^k = e$ is called the *order of g*.

Exercise 1.53 As a subgroup of S_4, we have

$$D_4 = \{(1),\ (1,3),\ (2,4),\ (1,2)(3,4),\ (1,4)(2,3),\ (1,2,3,4),\ (1,3)(2,4),\ (1,4,3,2)\}.$$

Verify both geometrically and computationally that D_4 is closed, and that each element in D_4 has an inverse that is also in D_4. Each of the reflections, along with the identity, (for example, $\{(1),(1,3)\}$) constitutes a subgroup of order 2, and the rotations form a subgroup of order 4 that contains a subgroup of order 2.

Exercise 1.54 The key observation here is that any two-cycle is its own inverse.

Exercise 1.55 Writing out the 6 subgroups of S_3 should be routine. I was just kidding about the others: There are 30 subgroups of S_4 (you should give this one a try), and 156 subgroups of S_5. There are 1455 subgroups of S_6.

You should try and get a feel for what some of the subgroups of S_4 and S_5 "look like." For example, the subgroups of S_5 that permute only three letters are "just like" S_3. How many such subgroups are there?

Exercise 1.56 Use the properties of the determinant function. For example, if A and B are in $SL(n, \mathbb{R})$ then $\mathrm{Det}(AB^{-1}) = \mathrm{Det}(A)\,\mathrm{Det}(B)^{-1} = 1 \cdot 1 = 1$, hence AB^{-1} is in $SL(n, \mathbb{R})$, and thus $SL(n, \mathbb{R})$ is a subgroup of $GL(n, \mathbb{R})$ by Theorem 1.48.

Exercise 1.57 Apply the definitions involved and the properties of the matrix transpose. For example, If A and B are in $O(n, \mathbb{R}) = O(n)$ then

$$
\begin{aligned}
(AB^{-1})^T (AB^{-1}) &= (B^{-T} A^T)(AB^{-1}) \quad \text{(using properties of transpose)} \\
&= B^{-T} (A^T A) B^{-1} \quad \text{(by associativity)} \\
&= B^{-T} I B^{-1} \qquad\quad \text{(because } A \in O(n)) \\
&= B^{-T} B^{-1} \\
&= (B B^T)^{-1} \\
&= I \qquad\qquad\qquad \text{(because } B \in O(n)).
\end{aligned}
$$

Hence $AB^{-1} \in O(n)$ and so $O(n)$ is a subgroup by the One Step test. It is a standard fact from linear algebra that orthogonal matrices preserve the standard inner product on $\mathbb{R}^n$, and thereby preserve lengths and angles. It follows that the orbit of any point $s \in \mathbb{R}^n$ is the set of all points lying the same distance as s from the origin. That is, a hyper-sphere in $\mathbb{R}^n$.

The group $O(n, \mathbf{R})$ is often referred to as the *group of rigid motions* in $\mathbb{R}^n$, that is, rotations and reflections. It is routine to check that if $A \in O(n, \mathbf{R})$, then $\mathrm{Det}\,A = \pm 1$. The subgroup $SO(n, \mathbf{R}) = \{A \in O(n, \mathbf{R}), \mathrm{Det} = 1\}$, called the *special orthogonal group*, preserves orientations, and so is the group of rotations in $\mathbb{R}^n$.

Exercise 1.64 Let I denote the identity matrix, and let E_{ij} denote an elementary matrix (with a 1 in the i, j position, and zeroes elsewhere) so that $I + E_{ij} \in GL(n, \mathbb{C})$. If M is central, then $(I + E_{ij})M = M(I + E_{ij})$, and hence $E_{ij}M = ME_{ij}$.

Now the ME_{ij} consists just one non-zero column, the ith column of M is the jth column of the product (and zeroes elsewhere), while $E_{ij}M$ consists just one non-zero row, the jth row of M is the ith row of the product (and zeroes elsewhere). For these two products to be equal, we must have $M_{i,i} = M_{j,j}$ for all i, j. Therefore, M must be a scalar multiple of the identity matrix I.

Exercise 1.74 Here is the verification of Example 1.72. Let $t, s \in \mathbb{R}$, and define

$$\phi : (\mathbb{R}, +) \to SL(2, \mathbb{R}) \quad \text{by} \quad \phi(t) = \begin{pmatrix} 1 & t \\ 0 & 1 \end{pmatrix}.$$

Then

$$\phi(t)\phi(s) = \begin{pmatrix} 1 & t \\ 0 & 1 \end{pmatrix}\begin{pmatrix} 1 & s \\ 0 & 1 \end{pmatrix} = \begin{pmatrix} 1 & t+s \\ 0 & 1 \end{pmatrix} = \phi(t+s).$$

Note the two different operations of matrix multiplication and addition of real numbers.

Exercise 1.75 Let G be the set of all permutations in S_n that fix some $k \in \{1, \dots, n\}$. Verifying that G is a subgroup is a routine application of the One Step Test for Subgroups. The groups G and S_{n-1} are isomorphic because they are permutation groups on a set with $n - 1$ elements. The actual "labels" of the elements being permuted is immaterial.

Exercise 1.78 Since $n \geq 3$, we can find positive integers i, j, k such that $1 \leq i < j < k \leq n$. Then check that the transpositions (i, j) and (j, k) don't commute. Alternatively, for $n \geq 3$, every group S_n contains S_3 as a subgroup, which we can verify directly is non-Abelian.

Definition 1.87 We'll take this opportunity for a review. Suppose V and W are finite-dimensional vector spaces, and let $\phi : V \to W$ be linear. Choose a basis $\{v_1, \dots, v_m\}$ for V, and a basis $\{w_1, \dots, w_n\}$ for W. Then, if

$$\mathbf{v} = r_1\mathbf{v}_1 + \cdots r_m\mathbf{v}_m, \quad \text{and if} \quad \phi(\mathbf{v}) = \mathbf{w} = s_1\mathbf{w}_1 + \cdots s_n\mathbf{w}_n, \quad \text{for scalars } r_1, \dots, s_n,$$

then there is an $n \times m$ matrix $\tilde{\phi}$, called the *matrix realization of* ϕ, such that

$$
\begin{bmatrix} s_1 \\ s_2 \\ \vdots \\ s_n \end{bmatrix} = \begin{bmatrix} & & \\ & \tilde{\phi} & \\ & & \end{bmatrix} \begin{bmatrix} r_1 \\ r_2 \\ \vdots \\ r_m \end{bmatrix},
$$

and where the ith column of $\tilde{\phi}$ is the coordinate vector of $\phi(\mathbf{v}_i)$ with respect to the basis in W. It is customary to write the coordinate vectors as column vectors so that we can multiply by matrices on the left, just as we write $f(x)$ rather that $(x)f$.

Exercise 1.86 For $\sigma \in S_3$, let $\tilde{\sigma}$ denote the matrix associated with the permutation representation on $\mathbb{R}^3$. For example, using the hint we have

$$
\widetilde{(1,2)} = \begin{pmatrix} 0 & 1 & 0 \\ 1 & 0 & 0 \\ 0 & 0 & 1 \end{pmatrix} \quad \text{and} \quad \widetilde{(1,2,3)} = \begin{pmatrix} 0 & 0 & 1 \\ 1 & 0 & 0 \\ 0 & 1 & 0 \end{pmatrix}.
$$

Check that $\widetilde{\sigma\tau} = \tilde{\sigma}\,\tilde{\tau}$, and thus the map $\sigma \mapsto \tilde{\sigma}$ is a homomorphism. More generally, we know from linear algebra that composition of linear maps corresponds to multiplication of their respective matrix realizations.

Exercise 1.88

$$
\tilde{a} = \begin{pmatrix} 1 & 0 & 0 \\ 0 & a^{-1} & 0 \\ 0 & 0 & a^{-2} \end{pmatrix}.
$$

Exercise 1.90 Here are a few example computations to get you started. Position the triangle so that the origin is at the triangle's center, and so that a vertex is on the y-axis.

Reflection across the line running from the top vertex to the midpoint of the base, *i.e.* the y-axis, has the following action on the standard basis vectors

$$
\mathbf{e}_1 = \begin{bmatrix} 1 \\ 0 \end{bmatrix} \longmapsto \begin{bmatrix} -1 \\ 0 \end{bmatrix}, \quad \text{and} \quad \mathbf{e}_2 = \begin{bmatrix} 0 \\ 1 \end{bmatrix} \longmapsto \begin{bmatrix} 0 \\ 1 \end{bmatrix},
$$

so the matrix realization of this symmetry is

$$
\begin{pmatrix} -1 & 0 \\ 0 & 1 \end{pmatrix}.
$$

For a counterclockwise rotation by $60°$ we have

$$\mathbf{e}_1 \longmapsto \begin{bmatrix} \cos\frac{\pi}{3} \\ \sin\frac{\pi}{3} \end{bmatrix} = \begin{bmatrix} \frac{1}{2} \\ \frac{\sqrt{3}}{2} \end{bmatrix}, \quad \text{and} \quad \mathbf{e}_2 \longmapsto \begin{bmatrix} -\sin\frac{\pi}{3} \\ \cos\frac{\pi}{3} \end{bmatrix} = \begin{bmatrix} -\frac{\sqrt{3}}{2} \\ \frac{1}{2} \end{bmatrix},$$

so the matrix realization of this symmetry is

$$\begin{pmatrix} \frac{1}{2} & -\frac{\sqrt{3}}{2} \\ \frac{\sqrt{3}}{2} & \frac{1}{2} \end{pmatrix}.$$

The matrix realizations for the other rotations and reflections are determined similarly.

Exercise 1.94 For $\mathbf{v} = v_1\mathbf{e}_1 + \cdots + v_n\mathbf{e}_n$, and for $f \in V^*$, let $a_i = f(\mathbf{e}_i) = a_i\mathbf{e}_i^*(\mathbf{e}_i)$. Then

$$f(\mathbf{v}) = f(v_1\mathbf{e}_1 + \cdots + v_n) = v_1 f(\mathbf{e}_1) + \cdots + v_n f(\mathbf{e}_n) = a_1 v_1 + \cdots + a_n v_n, \quad \text{so } f = a_1\mathbf{e}_1^* + \cdots + a_n\mathbf{e}_n^*.$$

Therefore, the dual basis spans V^*.

To show linear independence, suppose that $f = a_1\mathbf{e}_1^* + \cdots + a_n\mathbf{e}_n^* = 0$. But then $f(\mathbf{e}_i) = a_i = 0$ for all i.

Exercise 1.95 Let $\mathbf{e}_i^*$ be a vector in the dual basis. Then

$$\mathbf{e}_i^*\mathbf{v} = \mathbf{e}_i^*(r_1\mathbf{e}_1 + \cdots + r_i\mathbf{e}_i + \cdots + r_n\mathbf{e}_n)$$
$$= r_1\mathbf{e}_i^*(\mathbf{e}_1) + \cdots + r_i\mathbf{e}_i^*(\mathbf{e}_i) + \cdots + r_n\mathbf{e}_i^*(\mathbf{e}_n)$$
$$= r_1 \cdot 0 + \cdots r_i \cdot 1 + \cdots r_n \cdot 0$$
$$= r_i$$
$$= x_i(\mathbf{v}), \quad \text{the } i\text{th coordinate function on } \mathbf{v}.$$

Exercise 1.97 For place-permutation actions we start with n items labeled by $1, \ldots, n$, such as $x_1 x_2 \ldots x_n$, but where position matters. There are two possible ways for $\sigma \in S_n$ to act by place-permutation.

(1) An element σ can move the item in the $\sigma(j)$th position to the jth position. In this case the action is given by $\sigma.x_j = x_{\sigma(j)}$. It is important to remember that permutations act on <u>positions</u>. For example, if $\sigma = (1, 2, 3)$ and $\tau = (2, 3, 4)$, then

$$\tau.[\sigma.x_1 x_2 x_3 x_4] = (2, 3, 4).[(1, 2, 3).x_1 x_2 x_3 x_4] = (2, 3, 4).x_2 x_3 x_1 x_4 = x_2 x_1 x_4 x_3,$$

because the permutation $(2, 3, 4)$ moves the entry now in position 3, namely x_1, to position 2, and moves the entry in position 2, namely x_3, to position 4, etc.

Another way to see this is to relabel the result $x_2 x_3 x_1 x_4$ by position; set the entry in the first position, namely x_2, equal to y_1, set the entry in the second position, namely x_3, equal to y_2, etc. That is,

$$\sigma.x_1 x_2 x_3 x_4 = x_2 x_3 x_1 x_4 = y_1 y_2 y_3 y_4,$$

and consequently

$$\tau.y_1 y_2 y_3 y_4 = y_{\tau(1)} y_{\tau(2)} y_{\tau(3)} y_{\tau(4)} = y_1 y_3 y_4 y_2 = x_2 x_1 x_4 x_3$$

as before. Note that $y_j = \sigma.x_j = x_{\sigma(j)}$, so that

$$\tau.[\sigma.x_j] = \tau.y_j = y_{\tau(j)} = x_{\sigma(\tau(j))} = [\sigma\tau].x_j.$$

In this case the action is said to be *contravariant* since it reverses the order of τ and σ.

(2) An element σ can move the item in the jth position to the $\sigma(j)$th position, in which case the group action is given by $\sigma.x_j = x_{\sigma^{-1}(j)}$. Again, permutations act on positions. For example,

$$(2, 3, 4).[(1, 2, 3).x_1 x_2 x_3 x_4 x_5] = (2, 3, 4).x_3 x_1 x_2 x_4 x_5 = x_3 x_4 x_1 x_2 x_5.$$

In this case $\tau.[\sigma.x_j] = \sigma.x_{\tau^{-1}(j)} = x_{\sigma^{-1}(\tau^{-1}(j))} = x_{(\tau\sigma)^{-1}(j)} = \tau\sigma.x_j$, and the action is said to be *covariant* since it preserves the order of τ and σ.

Polynomials, Subspaces and Subrepresentations

2

The symmetric group S_n acts on the vector space of polynomials in n variables or indeterminants by permuting them in an obvious way. This representation provides easy-to-understand examples of some of the basic concepts in representation theory, including subrepresentations and intertwining maps.

2.1 Polynomials

A polynomial p is a formal linear combination of the form

$$p = C_{(\alpha_1,\ldots,\alpha_n)} x_1^{\alpha_1} x_2^{\alpha_2} \cdots x_n^{\alpha_n} + C_{(\beta_1,\ldots,\beta_n)} x_1^{\beta_1} x_2^{\beta_2} \cdots x_n^{\beta_n} + \cdots + C_{(\gamma_1,\ldots,\gamma_n)} x_1^{\gamma_1} x_2^{\gamma_2} \cdots x_n^{\gamma_n},$$

where $x_1, \ldots, x_n$ are *indeterminants* (that is, fixed symbols without any value), the exponents and subscripts $\alpha_1, \ldots, \gamma_n$ are non-negative integers, and the *coefficients*

$$C_{(\alpha_1,\ldots,\alpha_n)}, C_{(\beta_1,\ldots,\beta_n)}, \ldots, C_{(\gamma_1,\ldots,\gamma_n)}$$

are scalars in $\mathbb{C}$ (although they can be from any field). When necessary or convenient we can write this more compactly as

$$p = \sum C_\alpha x^\alpha,$$

where $\alpha := (\alpha_1, \ldots, \alpha_n)$ is a multi-index, sometimes called the *exponent sequence*, and $x^\alpha := x_1^{\alpha_1} x_2^{\alpha_2} \cdots x_n^{\alpha_n}$. The *degree* of the monomial x^α is $|\alpha| := \alpha_1 + \alpha_2 + \cdots + \alpha_n$, and the *total degree* of any polynomial p is the largest degree of any monomial appearing as a summand in p.

Any polynomial p can define a *polynomial function* $p(\mathbf{x})$ on an n-dimensional vector space V, where now $\mathbf{x} = (x_1, \ldots, x_n)$ is the argument of a function, *i.e.*, the vector of coordinate functions on V (Exercise 1.95), and the x_i are called *variables*.

© The Author(s), under exclusive license to Springer Nature Switzerland AG 2022
R. M. Howe, *An Invitation to Representation Theory*, SUMS Readings,
https://doi.org/10.1007/978-3-030-98025-2_2

For example, if $V = \mathbb{R}^2$, if $p(x_1, x_2) = x_1^2 + x_2$, and $\mathbf{v} = (3, 4) \in \mathbb{R}^2$, then $p(\mathbf{v}) = 3^2 + 4$ since $x_1(\mathbf{v}) = 3$ and $x_2(\mathbf{v}) = 4$.

For our purposes it is probably easiest to just think of polynomials as formal linear combinations, but we will have occasion to exploit their incarnation as functions. It is common to use the notation $p(\mathbf{x})$ for a polynomial in either case.

Denote the collection of polynomials by $\mathcal{P}(\mathbf{x})$, $\mathcal{P}(x_1, \ldots, x_n)$, $\mathcal{P}(V)$, or just $\mathcal{P}$ if the vector space V or the number of indeterminants/variables is understood from the context (or is immaterial). Under the usual operations of polynomial addition and scalar multiplication, $\mathcal{P}$ is itself a vector space, but with the additional operation of polynomial multiplication. We call such an algebraic structure (a vector space with the additional operation of vector multiplication) an *algebra*.[1] We will usually write the polynomials in boldface when we wish to emphasize that we are treating them as vectors.

We define an action of S_n on $\mathcal{P}$ by in the obvious way. First we define the action on each indeterminate;

$$\sigma.\mathbf{x}_i := \mathbf{x}_{\sigma(i)},$$

then extend this definition multiplicatively, defining the action on the monomials by

$$\sigma.\mathbf{x}^\alpha = \sigma.(\mathbf{x}_1^{\alpha_1}\mathbf{x}_2^{\alpha_2}\cdots\mathbf{x}_n^{\alpha_n}) := \mathbf{x}_{\sigma(1)}^{\alpha_1}\mathbf{x}_{\sigma(2)}^{\alpha_2}\cdots\mathbf{x}_{\sigma(n)}^{\alpha_n},$$

and then extend by linearity,

$$\sigma.p := \sum C_\alpha\, \sigma.\mathbf{x}^\alpha.$$

This coincides with the action of S_n on the coordinate functions on V, as in Example 1.96, and therefore to the polynomial functions on V. For $n \leq 4$, it is usually less cumbersome to use $\mathbf{x}$, $\mathbf{y}$, $\mathbf{z}$, and $\mathbf{w}$ for the variables. When we wish to discuss something in full generality, we may need subscripted variables.

Example 2.1 Using the above definitions and conventions, we have

$$(1, 2, 3).(\mathbf{xy}^2\mathbf{z}^3 + 5\mathbf{x}^4\mathbf{y}^5\mathbf{z}^6) = \mathbf{yz}^2\mathbf{x}^3 + 5\mathbf{y}^4\mathbf{z}^5\mathbf{x}^6 = \mathbf{x}^3\mathbf{yz}^2 + 5\mathbf{x}^6\mathbf{y}^4\mathbf{z}^5 \text{ for } (1, 2, 3) \in S_3,$$

and

$$(1, 3, 2, 4).(\mathbf{xy}^2\mathbf{z}^3\mathbf{w}^4) = \mathbf{zw}^2\mathbf{y}^3\mathbf{x}^4 = \mathbf{x}^4\mathbf{y}^3\mathbf{zw}^2 \text{ for } (1, 3, 2, 4) \in S_4.$$

Remark 2.2 There are several ways to generalize polynomials.

---

[1] In the case of polynomials as formal linear combinations, the notation $\mathbb{F}[x_1, \ldots, x_n]$ is also common. Sometimes we see the terminology $\mathcal{P} = \mathcal{S}(V^*)$, the algebra of *symmetric tensors on the dual space* when considering them as functions.

2.2 Subspaces and Subrepresentations

For each non-negative integer k, denote by $\mathcal{P}_k \subseteq \mathcal{P}$ the *homogeneous polynomials of degree k*. The space $\mathcal{P}_k$ is the set of all polynomials where each term has degree exactly k, along with the zero polynomial. For example, $\mathbf{x}^2\mathbf{y} + \mathbf{z}^3 \in \mathcal{P}_3$ and $\mathbf{xyz}^2 + 2\mathbf{x}^2\mathbf{y}^2 \in \mathcal{P}_4$.

A homogeneous polynomial defines a homogeneous function; if t is any scalar, then p is homogeneous of degree k if

$$p(t\,\mathbf{x}_1, \ldots, t\,\mathbf{x}_n) = t^k\,p\,(\mathbf{x}_1, \ldots, \mathbf{x}_n).$$

Exercise* 2.3 Confirm that $\mathcal{P}_k$ is a subspace of $\mathcal{P}$. What is an obvious basis for $\mathcal{P}_k$? What is the dimension of $\mathcal{P}_k$? Describe the space $\mathcal{P}_0$.

Exercise 2.4 Choose some monomial $\mathbf{x}^\alpha \in \mathcal{P}(\mathbf{x}_1, \ldots, \mathbf{x}_n)$. What is the S_n-orbit of $\mathbf{x}^\alpha$? Do this for several different choices of $\mathbf{x}^\alpha$ and n.

It is easy to see that permuting the variables of any polynomial in $\mathcal{P}_k$ will not change its degree. In other words, if $\sigma \in S_n$, and if $p \in \mathcal{P}_k$, then $\sigma.p \in \mathcal{P}_k$, and therefore $\mathcal{P}_k$ is itself a representation of S_n. We call such a subspace a *G-invariant subspace*, or a *subrepresentation of G*. More precisely,

Definition 2.5 Let V be a representation of G, and let W be a subspace of V. If $g.\mathbf{w} \in W$ for all $g \in G$ and all $\mathbf{w} \in W$, then W is a *G-invariant subspace* or a *subrepresentation* of G.

2.3 Partitions and More Subrepresentations

The space of homogeneous polynomials of degree k has as a basis those monomials whose exponents sum to k. This notion generalizes and has further applications.

For a positive integer k, a *composition* of k is a way of writing k as the sum of a sequence of non-negative integers $\lambda := (\lambda_1, \lambda_2, \ldots, \lambda_\ell)$, with $|\lambda| := \lambda_1 + \lambda_2 + \ldots + \lambda_\ell = k$. For example, $(1, 2, 2)$, $(3, 2)$, and $(3, 2, 0, 0)$ are all compositions of 5. In this case we write $\lambda \vDash k$, and two compositions are equivalent if they are the same up to the order of the terms.

If we further require that $\lambda_1 \geq \lambda_2 \geq \ldots \geq \lambda_\ell \geq 0$, then we say that λ is a *partition of k with length ℓ*, and write $\lambda \vdash k$. Each partition $\lambda \vdash k$ is a representative of those compositions of k that are equivalent to λ. Of course, a partition may end in one or more zeros, and we will add or omit trailing zeros when necessary or convenient (which will change the length). Also, we may occasionally use the more specific term *integer partition* since there are other partitions in mathematics (partitions of sets, of intervals, etc.). Compositions, and especially partitions, will arise frequently when discussing the symmetric group and its representations.

Exercise* 2.6 Write out all of the partitions of k for several values of k. Hint: Don't be too ambitious; there are 42 partitions of 10.

Notation 2.7 The partition $(1, 1, \ldots, 1) \vdash n$ is often written (1^n).

As an immediate application, we will further decompose $\mathcal{P}_k$ into subspaces spanned by monomials of a particular "type" that are labeled by partitions of k. If the monomial $\mathbf{x}^\beta = \mathbf{x}_1^{\beta_1} \mathbf{x}_2^{\beta_2} \cdots \mathbf{x}_n^{\beta_n}$ is in $\mathcal{P}_k$, then $\beta \models k$, and we denote by $\mathcal{P}_\alpha$ the subspace of $\mathcal{P}_k$ spanned by all monomials $\mathbf{x}^\beta$ with the composition β equivalent to the partition $\alpha \vdash k$. For example:

$$\mathbf{x}^2\mathbf{y} \text{ and } \mathbf{x}\mathbf{y}^2 \text{ are in } \mathcal{P}_{(2,1)}(\mathbf{x}, \mathbf{y}),$$

$$\mathbf{x}^2\mathbf{y} \text{ and } \mathbf{x}\mathbf{y}^2 \text{ are in } \mathcal{P}_{(2,1,0)}(\mathbf{x}, \mathbf{y}, \mathbf{z}),$$

$$\mathbf{x}^2\mathbf{y}^2\mathbf{z} \text{ and } \mathbf{x}\mathbf{y}^2\mathbf{w}^2 \text{ are in } \mathcal{P}_{(2,2,1,0)}(\mathbf{x}, \mathbf{y}, \mathbf{z}, \mathbf{w}).$$

We call $\mathcal{P}_\alpha$ the *subspace of polynomials of type α*.

Exercise* 2.8 Verify that for each $\alpha \vdash k$, $\mathcal{P}_\alpha$ is a subspace of $\mathcal{P}_k(\mathbf{x}_1, \mathbf{x}_2, \ldots, \mathbf{x}_n)$. Determine a basis for some examples of $\mathcal{P}_\alpha$ for, say, $n \leq 4$ and $k \leq 5$.

It should be readily apparent that if we permute a monomial in $\mathcal{P}_\alpha$, we could never obtain a monomial of a different type. For example, there is no $\sigma \in S_3$ such that $\sigma.\mathbf{x}^2\mathbf{y} = \mathbf{x}^3$.

Proposition 2.9 *For each $\alpha \vdash k$, $\mathcal{P}_\alpha$ is an S_n-invariant subspace of $\mathcal{P}_k(\mathbf{x}_1, \mathbf{x}_2, \ldots, \mathbf{x}_n)$.*

Exercise 2.10 Prove the above proposition. While the result is intuitively obvious, you should work out the notation required to prove the general case.

Exercise 2.11 Choose a positive integer n, a partition $\alpha \vdash k$ for some k, and fix a monomial $\mathbf{x}^\alpha$. Check that the S_n-orbit of $\mathbf{x}^\alpha$ is a basis for the subspace $\mathcal{P}_\alpha$.

Another way of labeling partitions, and which has applications to polynomial spaces, is by *cycle class*.[2] If $\alpha \vdash k$ has length ℓ, we let m_0 denote the number of zeros in α, m_1 denote the number of ones, m_2 the number of twos, etc. The cycle class is then written $(0^{m_0}, 1^{m_1}, 2^{m_2}, \ldots, i^{m_i})$. For example, the partition $\alpha = (5, 3, 3, 2, 2, 2, 0, 0) \vdash 17$ has cycle class $(0^2, 1^0, 2^3, 3^2, 4^0, 5^1)$.

We will refer to the partition equivalent to the composition $m := (m_0, m_1, m_2, \ldots, m_i) \vdash \ell$ as the *signature* of the partition α. For the above

---

[2] The term arises when discussing the structure of S_n.

example, the partition α has cycle class $(0^2, 1^0, 2^3, 3^2, 4^0, 5^1)$ and signature $(3, 2, 2, 1, 0, 0) \vdash 8$, which is the partition equivalent to the composition $(2, 0, 3, 2, 0, 1)$. We extend this notion to polynomial spaces; we say that the space $\mathcal{P}_\alpha$ has signature m if α has signature m.

Remark 2.12 We emphasize that if S_n acts on $\mathcal{P}_\alpha \subseteq \mathcal{P}_k \subseteq \mathcal{P}(x_1, \ldots, x_n)$, and if α has signature $m = (m_0, m_1, m_2, \ldots, m_i)$, then $m \vdash n$ and $\alpha \vdash k$. For example, S_8 acts on $\mathcal{P}_{(5,3,3,2,2,2,0,0)} \subseteq \mathcal{P}_{17}$, and has signature $(3, 2, 2, 1, 0, 0) \vdash 8$. We will use this idea when we discuss Young permutation modules in Chap. 10.

Exercise* 2.13 Compute the dimension of $\mathcal{P}_\alpha(x_1, \ldots, x_n)$ for several cases of α and n. Find a general formula for the dimension of $\mathcal{P}_\alpha(V)$.

2.4 Vector Space Direct Sums

Here is another instance of "busting things up into sub-things," and conversely, "pasting them together."

Definition 2.14 If V is a vector space and if U and W are subspaces of V, we say that V is the *(internal) direct sum* of U and W if :

(1) $V = U + W$. That is, any $\mathbf{v} \in V$ can be written as $\mathbf{v} = \mathbf{u} + \mathbf{w}$ for some $\mathbf{u} \in U$ and some $\mathbf{w} \in W$; and
(2) $U \cap W = \{0\}$.

In this case we write $V = U \oplus W$, and say that U and W are *direct summands* of V.

More generally, given a vector space V with subspaces $V_1, V_2, \ldots, V_k$, then

(1$'$) the *sum*

$$\sum_{i=1}^{k} V_i = V_1 + V_2 + \cdots + V_k := \{\mathbf{v}_1 + \mathbf{v}_2 + \cdots + \mathbf{v}_k \mid \mathbf{v}_i \in V_i\}$$

is a subspace of V (check this).
(2$'$) The sum is a *direct sum*, written $\bigoplus_{i=1}^{k} V_i$, if

$$V_r \cap \sum_{i \neq r} V_i = \{0\} \quad \text{for } r = 1, \ldots, k.$$

Exercise* 2.15 Show that if $V = U \oplus W$, then the sum $\mathbf{v} = \mathbf{u} + \mathbf{w}$ as in Definition 2.14 (1) above is unique. This is an important property of direct sums. Show that this uniqueness condition and the condition that $U \cap W = \{0\}$ are equivalent.

Exercise 2.16 Let $V = \bigoplus_{i=1}^{k} V_i$, and for each i choose a vector $\mathbf{v}_i \in V_i$. Show that the set $\{\mathbf{v}_i\}$ is a linearly independent subset of V. Hint: Induction on k.

Exercise 2.17 Let $V = \mathbb{R}^3$. Let V_1 be the subspace spanned by the vector $(1, 0, 0)$, let V_2 be the subspace spanned by the vector $(0, 1, 0)$, and let V_3 be the subspace spanned by the vector $(1, 1, 0)$. Check that $V_1 \cap V_2 = V_1 \cap V_3 = V_2 \cap V_3 = \{0\}$, but the sum $V_1 + V_2 + V_3$ is not direct. Conclude that condition (2′) above is required for the sum to be direct, and that the condition

$$V_i \cap V_j = \{0\} \text{ for } i \neq j$$

is not sufficient.

Example 2.18 The most obvious examples of direct sums of vector spaces are the Cartesian plane $\mathbb{R}^2 = \mathbb{R} \oplus \mathbb{R}$, and 3-space $\mathbb{R}^3 = \mathbb{R} \oplus \mathbb{R} \oplus \mathbb{R} = \mathbb{R}^2 \oplus \mathbb{R}$.

On the other hand, given any two (or more) vector spaces U and W, we can form the *(external) direct sum* by defining $V = U \oplus W$ as the set of ordered pairs $(\mathbf{u}, \mathbf{w})$ with $\mathbf{u} \in U$ and $\mathbf{w} \in W$, and with addition and scalar multiplication defined componentwise:

$$(\mathbf{u}, \mathbf{w}) + (\mathbf{u}', \mathbf{w}') = (\mathbf{u} + \mathbf{u}', \mathbf{w} + \mathbf{w}'), \quad \text{and} \quad r(\mathbf{u}, \mathbf{w}) = (r\mathbf{u}, r\mathbf{w}).$$

The space U (say) can then be considered as a direct summand of V by the identification $U \cong \{(\mathbf{u}, 0) \mid \mathbf{u} \in U\}$. Since the distinction is typically clear from the context or immaterial, we will usually drop the designation "internal" or "external." By the way, there are notions of direct sums for other mathematical structures such as rings and modules.

Remark 2.19 There can be lots of ways to decompose a vector space as a direct sum of its subspaces. For example, if $V = \mathbb{R}^2$, and if U and W are the subspaces spanned by the vectors $(1, 0)$ and $(0, 1)$ respectively, then $V = U \oplus W$. But we can also take U' and W' to be the subspaces spanned by the vectors $(1, 0)$ and $(1, 1)$ respectively, and again $V = U' \oplus W'$. Decomposing vector spaces into direct sums in some "natural" or "useful" way is often an important problem. Here is an example familiar from basic physics: when determining frictional forces acting on a mass placed on an inclined surface, it is useful to resolve the downward gravitational force on the mass into two components; one parallel to, and one normal to the surface.

Exercise* 2.20 Show that if V is a finite dimensional vector space, and if $V = U \oplus W$, then $\text{Dim } V = \text{Dim } U + \text{Dim } W$.

Exercise 2.21 Let V be a vector space with basis $\{\mathbf{v}_1, \mathbf{v}_2, \ldots, \mathbf{v}_n\}$, and let V_i be the one-dimensional subspace of V spanned by $\mathbf{v}_i$. Show that $V = \bigoplus_{i=1}^{n} V_i$.

Exercise 2.22 Show that $\mathcal{P}_k = \oplus_{\mu \vdash k} \mathcal{P}_\mu$. The notation means that the direct sum is taken over all partitions of k. This is intuitively obvious, but it is a good exercise to work out the notation and write out the details.

Exercise* 2.23 Show that if V is a representation of G, and if U and W are G-invariant subspaces of V, then $U \oplus W$ is a G-invariant subspace of V. That is, the direct sum of two (or more) representations of G is also a representation of G.

Remark* 2.24 Since any polynomial can be written as a finite sum of homogeneous polynomials, we can write $\mathcal{P} = \oplus_{k=0}^{\infty} \mathcal{P}_k$. Additionally, if $p \in \mathcal{P}_k$, and if $q \in \mathcal{P}_\ell$, then the product pq lies in $\mathcal{P}_{k+\ell}$, so $\mathcal{P}$ is said to have the structure of a *graded algebra*, or more specifically, a $\mathbb{Z}$-graded algebra (we set $\mathcal{P}_k = \{0\}$ whenever $k < 0$). Moreover, since the action of S_n does not change the degree, this action of S_n is a *graded algebra homomorphism*.

2.5   Projection Maps

If $V = U \oplus W$, and if $\mathbf{v} = \mathbf{u} + \mathbf{w}$ as in Exercise 2.15, then the map $P_U : V \to U$ given by $P_U(\mathbf{v}) = P_U(\mathbf{u} + \mathbf{w}) = \mathbf{u}$ is called the *projection onto U along W*. In general, a linear map $P : V \to V$ is a projection if the composition $P \circ P = P$.

Exercise 2.25 Find more examples of projection maps.

Exercise* 2.26 Verify the assertions implicit in the above statement; the map P is well-defined, linear, onto U, and that $P \circ P = P$.

Exercise* 2.27 Show that if $V = U \oplus W$, and if P_U is the projection onto U along W, then $\operatorname{Ker} P_U = W$. Generalize this result: If $P : V \to V$ is linear, and if $P \circ P = P$, then $V = \operatorname{Im} P \oplus \operatorname{Ker} P$.

Exercise 2.28 Let V be a representation of a finite group G. Show that the set $V^G := \{\mathbf{v} \in V \mid g.\mathbf{v} = \mathbf{v} \text{ for all } g \in G\}$, that is, the set of vectors fixed by G, is an invariant subspace of V. Show that the map $\phi : V \to V$ given by

$$\phi(\mathbf{v}) := \frac{1}{|G|} \sum_{g \in G} g.\mathbf{v}$$

is linear and a projection onto V^G. Hint: This is a routine application of the definitions involved.

2.6　　Irreducible Subspaces

Yet another theme in mathematics is that, given some mathematical structure, is there some sort of "indecomposable" or "atomic" or "prime" substructure? A familiar example is writing any positive integer as a product of prime numbers.

We have seen that permuting the indeterminants in $\mathcal{P}(\mathbf{x_1}, \dots, \mathbf{x_n})$ is a representation of S_n. We also have that $\mathcal{P}_k$, the homogeneous polynomials of degree k, are S_n-invariant subspaces of $\mathcal{P}$. Furthermore, for each $\alpha \vdash k$, $\mathcal{P}_\alpha$ is an S_n-invariant subspace of $\mathcal{P}_k$. Can we decompose any arbitrary representation into a direct sum of smaller subrepresentations? Is such a decomposition unique? Can we keep going to obtain an unending chain of invariant subspaces or does this process have to end? We will see that the answer to these questions can be nuanced.

Example 2.29 Consider the permutation representation of S_3 on $\mathcal{P}_1 = \mathcal{P}_{(1,0,0)} =$ span of $\{\mathbf{x}, \mathbf{y}, \mathbf{z}\}$. Let $\mathcal{I}$ denote the subspace of all polynomials in $\mathcal{P}_1$ of the form $\{r\mathbf{x} + r\mathbf{y} + r\mathbf{z} \mid r$ any scalar$\}$. Let $\mathcal{W}$ denote the subspace of all polynomials in $\mathcal{P}_1$ of the form $\{r\mathbf{x} + s\mathbf{y} + t\mathbf{z} \mid r + s + t = 0\}$. Then $\mathcal{I}$ and $\mathcal{W}$ are S_3-invariant subspaces that have no proper invariant subspaces. Recall that a subspace W of a vector space V is *proper* if $W \neq \{0\}$ and $W \neq V$. This example motivates the following definition.

Definition 2.30 A representation of a group G on a vector space W is *irreducible* if W contains no proper G-invariant subspaces. You might occasionally see this term shortened to *irrep*, especially by physicists or those wishing to save ink. Writing "irrep" is certainly easier than writing out "irreducible representation."

Exercise* 2.31 Prove the assertions in Example 2.29, that $\mathcal{I}$ and $\mathcal{W}$ are subspaces and are irreducible. Find a basis for each of the subspaces $\mathcal{I}$ and $\mathcal{W}$. Show that $\mathcal{P}_1 = \mathcal{I} \oplus \mathcal{W}$.

Remark 2.32 The subspace $\mathcal{I}$ in Example 2.29 is an example of a trivial representation of S_3 (Definition 1.85). The subspace $\mathcal{W}$, complementary to $\mathcal{I}$ in the permutation representation, is often called the *standard representation*.

Exercise* 2.33 Work out examples analogous to Example 2.29 for the spaces $\mathcal{P}_{(2,0,0)}$, $\mathcal{P}_{(3,0,0)}$, and $\mathcal{P}_{(1,1,0)}$.

Exercise* 2.34 We will use these results in a later section. Let $\mathbf{w}_1 = \frac{1}{\sqrt{2}}(\mathbf{x} - \mathbf{y})$ and $\mathbf{w}_2 = \frac{1}{\sqrt{6}}(\mathbf{x} + \mathbf{y} - 2\mathbf{z})$ be two basis vectors for the two-dimensional standard representation $\mathcal{W}$ of S_3 in $\mathcal{P}_1(\mathbf{x}, \mathbf{y}, \mathbf{z})$. Write out the matrix realizations (Definition 1.87) for this representation.

That is, if $\mathbf{v} = a\mathbf{w}_1 + b\mathbf{w}_2$ for some scalars a, b, and if $\sigma \in S_n$, then $\sigma.\mathbf{v} = a'\mathbf{w}_1 + b'\mathbf{w}_2$ where the matrix realization $\tilde{\sigma}$ acts on the coordinate (column) vector by matrix multiplication from the left;

$$\tilde{\sigma} \begin{bmatrix} a \\ b \end{bmatrix} = \begin{bmatrix} a' \\ b' \end{bmatrix}.$$

Equivalently, you could consider the permutation representation of S_3 on $\mathbb{R}^3$ as in Example 1.22, where the irreducible two-dimensional subrepresentation has as a basis the vectors $\mathbf{w}_1 = \frac{1}{\sqrt{2}}(1, -1, 0)$ and $\mathbf{w}_2 = \frac{1}{\sqrt{6}}(1, 1, -2)$. Note that this basis is orthonormal with respect to the usual dot product on $\mathbb{R}^3$. Verify that the matrices you obtain are orthogonal (Exercise 1.57).

Exercise 2.35 Let V be a representation of a group G, and suppose that U and W are G-invariant subspaces of V. Show that $U \cap W$ is a G-invariant subspace of both U and W. Conclude that if U is irreducible, then either $U \cap W = \{0\}$ or U is a subrepresentation of G contained in W. It then follows that if both U and W are irreducible, then either $U \cap W = \{0\}$ or $U = W$. Hint: That $U \cap W$ is a vector subspace of both U and W is a standard exercise from linear algebra. That $U \cap W$ is G-invariant is a routine application of the definition.

The above examples illustrate some of the basic issues in representation theory. Given a representation of a group, we have seen that we may be able to find "smaller" subrepresentations of a given representation, and that some of these may contain no proper subrepresentations. A given representation may decompose into a direct sum of these irreducible subrepresentations. Conversely, given any two (or more) representations of a group we can "paste" them together by taking their direct sum (internal or external as needed).

This leads to two questions:

(1) What are the irreducible representations, and how do we describe them?
(2) Given any representation of a group G, can we bust it up into irreducible parts, and if so, how? What are these parts?

As is often the case, we may need to refine exactly what we mean by these questions:

(1') Can we find all of the irreducible representations (up to some sort of equivalence) of a given group? How do we know this is all of them? Is there some way to label the inequivalent irreducible representations? Is there some natural basis for each one?
(2') How does a given representation decompose into its irreducible parts? Can we always do this? If so, how?

2.7   Hints and Additional Comments

Remark 2.2 Generalizations of polynomials.

A *power series* is like a polynomial, but with infinitely many terms corresponding to the non-negative integers. For example

$$1 + 2x + 5x^2 + 3x^4 + \cdots .$$

A *formal power series* is one in which x is an indeterminate, not a variable,[3] so we can ignore any questions of convergence (absolute, conditional, uniform, or otherwise), and therefore we can add, subtract and multiply formal power series fearlessly.

However, if we treat the x as a variable, as with Taylor series, then we are forced to consider issues of convergence. For example,

$$1 + x^2 + x^3 + x^4 + \cdots = \frac{1}{1-x}$$

is true only for those values of x that lie in the interval $[-1, 1)$, but this is true formally when x is an indeterminate. You should try this, either using long division (as with polynomials), or by verifying that

$$1 = (1-x)(1 + x^2 + x^3 + x^4 + \cdots)$$

by multiplying out the right-hand side. Note that you can never stop!

A *Laurent polynomial* is like a polynomial, but allowing negative exponents. For example

$$x^{-3} + 3x^{-2} + 2 + x + 4x^2,$$

and this notion can be extended to *Laurent series*, both as formal series and as Laurent series expansions of functions.

A *matrix polynomial or power series* has square matrices as variables or indeterminates.

A *trigonometric polynomial* is a linear combination of the functions $\sin(nx)$ and $\cos(nx)$ such as

$$2 + 3\sin(x) + 4\cos(x) + 5\sin(2x) + 6\cos(2x) + 7\cos(3x).$$

There are also *trigonometric series*.

---

[3] There can be multiple indeterminants or variables.

Exercise 2.3 Verifying that $\mathcal{P}_k$ is a subspace of $\mathcal{P}$ is a routine application of the "Subspace Test" from linear algebra. The distinct monomials of degree k are an obvious and convenient basis for $\mathcal{P}_k$, and the dimension of $\mathcal{P}_k$ is then the number of distinct monomials of degree k in n variables. There is a combinatorial formula that provides this number, but here is a nifty "proof by picture."

We let k dots represent the degree of each monomial, which we then apportion among the n variables by dividing up the dots with $n - 1$ vertical lines. The figure below depicts this for the monomial $x^3 y^2 z^2$.

$$\cdots \mid \cdots \mid \cdots$$

Thus there are $\binom{n+k-1}{n-1}$ ways to position the $n - 1$ lines among the $n+k-1$ dots and lines, which is the dimension of the space of homogeneous polynomials of degree k in n variables. The space $\mathcal{P}_0$ consists of the constant polynomials.

Exercise 2.6 To get you started, here are the partitions of 5;

$$(5), \ (4, 1), \ (3, 2), \ (3, 1, 1), \ (2, 2, 1), \ (2, 1, 1, 1), \ \text{and} \ (1, 1, 1, 1, 1).$$

Exercise 2.8 An obvious basis for $\mathcal{P}_\alpha$ is to take all permutations of $\mathbf{x}^\alpha$. For example, the set $\mathcal{B} = \{\mathbf{x}^2\mathbf{y}, \mathbf{xy}^2, \mathbf{x}^2\mathbf{z}, \mathbf{xz}^2, \mathbf{y}^2\mathbf{z}, \mathbf{yz}^2\}$ is a basis for $\mathcal{P}_{(2,1,0)}(\mathbf{x}, \mathbf{y}, \mathbf{z})$. Note that $\mathcal{B}$ is the S_3-orbit of any one of its elements.

Exercise 2.13 Consider the subspace $\mathcal{P}_{(2,2,1,1,1,0)}(\mathbf{x}_1, \ldots, \mathbf{x}_6)$. There are 6! ways to permute the variables in, say, the monomial $\mathbf{x}_1^2\mathbf{x}_2^2\mathbf{x}_3\mathbf{x}_4\mathbf{x}_5$, but some of them have no effect; the 2! permutations of $\mathbf{x}_1$ and $\mathbf{x}_2$, and the 3! permutations of $\mathbf{x}_3$, $\mathbf{x}_4$ and $\mathbf{x}_5$. Thus there are $\frac{6!}{2!3!}$ distinct monomials in $\mathcal{P}_{(2,2,1,1,1,0)}(\mathbf{x}_1, \ldots, \mathbf{x}_6)$. More generally, if $\mathcal{P}_\mu(\mathbf{x}_1, \ldots, \mathbf{x}_n)$ has signature λ, then it's dimension is $\frac{n!}{\lambda!}$, where $\lambda! := \lambda_1!\lambda_2!, \ldots \lambda_n!$.

Exercise 2.15 To show that the sum is unique, we suppose that we can write $\mathbf{v}$ in two different ways;

$$\mathbf{v} = \mathbf{u} + \mathbf{w} = \mathbf{u}' + \mathbf{w}' \ \text{for some} \ \mathbf{u}, \mathbf{u}' \in U \ \text{and} \ \mathbf{w}, \mathbf{w}' \in W.$$

This implies that $\mathbf{u} - \mathbf{u}' = \mathbf{w}' - \mathbf{w} \in U \cap W = \{\mathbf{0}\}$. Thus $\mathbf{u} = \mathbf{u}'$ and $\mathbf{w} = \mathbf{w}'$.

If every $\mathbf{v} \in V$ can be written uniquely as $\mathbf{v} = \mathbf{u} + \mathbf{w}$ for some $\mathbf{u} \in U$ and some $\mathbf{w} \in W$, and if $\mathbf{v} \in U \cap W$, then $\mathbf{v} = \mathbf{u} + \mathbf{0} = \mathbf{0} + \mathbf{w}$, and hence $\mathbf{v} = \mathbf{u} = \mathbf{w} = \mathbf{0}$.

Exercise 2.20 This is a standard result from linear algebra. Let $V = U \oplus W$, let $\{\mathbf{u}_1, \ldots, \mathbf{u_k}\}$ be a basis for U, and let $\{\mathbf{w}_{k+1}, \ldots, \mathbf{w_n}\}$ be a basis for W. We claim that $B = \{\mathbf{u}_1, \ldots, \mathbf{u_k}, \mathbf{w}_{k+1}, \ldots, \mathbf{w_n}\}$ is a basis for V. Certainly B spans V since we can write any $\mathbf{v} \in V$ as $\mathbf{v} = \mathbf{u} + \mathbf{w}$. To show that B is linearly independent, we set $r_1\mathbf{u_1} + \cdots + r_k\mathbf{u_k} + \cdots + r_{k+1}\mathbf{w_{k+1}} + \cdots + r_n\mathbf{w_n} = \mathbf{0}$. Some simple arithmetic gives

$$r_1\mathbf{u_1} + \cdots + r_k\mathbf{u_k} + \cdots = -r_{k+1}\mathbf{w_{k+1}} - \cdots - r_n\mathbf{w_n} \in U \bigcap W = \{\mathbf{0}\}.$$

But then all of the coefficients must be 0 by the linear independence of the bases for both U and W. The result follows.

Exercise 2.23 If $\mathbf{v} \in U \oplus W$, then we can write $\mathbf{v} = \mathbf{u} + \mathbf{w}$ for some $\mathbf{u} \in U$ and some $\mathbf{w} \in W$. Since $g \in G$ acts linearly, we have $g.\mathbf{v} = g.(\mathbf{u} + \mathbf{w}) = g.\mathbf{u} + g.\mathbf{w} \in U \oplus W$ because both U and W are G-invariant. Hence $U \oplus W$ is a G-invariant subspace of V.

Remark 2.24 More generally, if G is an additive abelian group, and if $A = \bigoplus_{i \in G} A_i$, is an algebra with an associative product $*$, where $\{A_i\}_{i \in G}$ is a family of algebras indexed by G, and such that $A_i * A_j \subset A_{i+j}$, then A is called a G-graded algebra.

In our case, we can further refine the grading by writing the homogeneous spaces $\mathcal{P}_k$ as a direct sum of the subspaces $\mathcal{P}_\alpha$ for $\alpha \vdash k$, and note that the action of S_n is still a graded algebra homomorphism. What is the abelian group that labels the homogeneous subspaces? Hint: Suppose $\alpha \vdash k$, and $\beta \vdash \ell$. If $\mathbf{p} \in \mathcal{P}_\alpha$ and if $\mathbf{q} \in \mathcal{P}_\beta$ where is $\mathbf{pq}$?

If we first try an example, say for $\mathbf{x^2y} \in \mathcal{P}_{(2,1,0)}$ and $\mathbf{xyz} \in \mathcal{P}_{(1,1,1)}$, then

$$(\mathbf{x^2y})(\mathbf{xyz}) = \mathbf{x^3y^2z} \in \mathcal{P}_{(3,2,1)}, \text{ where } (3,2,1) = (2,1,0) + (1,1,1) \text{ added componentwise.}$$

Thus $\mathcal{P}(x_1, \ldots, x_n)$ is graded with respect to the group $(\mathbb{Z}^n, +)$, the group of n-tuples of integers with componentwise addition.

Exercise 2.26 Projection maps are well defined by Exercise 2.15. Projection maps are surjective because if $\mathbf{u} \in U$, then $\mathbf{u} = P_U(\mathbf{u})$.

Exercise 2.27 To show that $V = \text{Im } P + \text{Ker } P$, let V be a vector space, and suppose that $P \colon V \to V$ is linear, with $P \circ P = P$. For any $\mathbf{v} \in V$, we certainly have $\mathbf{v} = P(\mathbf{v}) + [\mathbf{v} - P(\mathbf{v})]$, and clearly $P(\mathbf{v}) \in \text{Im } P$, so we need to show that $[\mathbf{v} - P(\mathbf{v})] \in \text{Ker } P$. A direct application of the hypotheses yields

$$P([\mathbf{v} - P(\mathbf{v})]) = P(\mathbf{v}) - P(P(\mathbf{v})) = P(\mathbf{v}) - P(\mathbf{v}) = 0, \text{ as desired.}$$

To show that this sum is direct, we suppose that $\mathbf{w} \in \text{Im } P \bigcap \text{Ker } P$. This means that $\mathbf{w} = P(\mathbf{v})$ for some $\mathbf{v} \in V$, and that $P(\mathbf{w}) = 0$. Then

$$\mathbf{w} = P(\mathbf{v}) = P(P(\mathbf{v})) = P(\mathbf{w}) = 0.$$

Hence $\text{Im } P \bigcap \text{Ker } P = \{\mathbf{0}\}$.

Exercise 2.31 The subrepresentation $\mathcal{I}$ is clearly irreducible since it is one dimensional, and the polynomial $\mathbf{x} + \mathbf{y} + \mathbf{z}$ is a convenient basis. There are lots of possible bases for $\mathcal{W}$, one being $\{\mathbf{x} - \mathbf{y}, \mathbf{x} - \mathbf{z}\}$, another being being $\{\mathbf{x} - \mathbf{y}, \mathbf{x} + \mathbf{y} - 2\mathbf{z}\}$.

To show that $\mathcal{P}_1 = \mathcal{I} \oplus \mathcal{W}$, we first note that if $\mathbf{q} = a\mathbf{x} + b\mathbf{y} + c\mathbf{z}$ (for some scalars a, b, c) is in $\mathcal{I} \cap \mathcal{W}$, then we must have both $a = b = c$ and $a + b + c = 0$, and hence $\mathbf{q} = 0$. To conclude that $\mathcal{P}_1 = \mathcal{I} + \mathcal{W}$, we can write an arbitrary polynomial in $\mathcal{P}_1$ in terms of the bases for $\mathcal{I}$ and $\mathcal{W}$, but it suffices to just add up the dimensions.

To show that $\mathcal{W}$ is irreducible, first note that any proper invariant subspace must be one-dimensional, and so spanned by a non-zero vector $\mathbf{p} = a\mathbf{x} + b\mathbf{y} + c\mathbf{z}$ for some scalars a, b, c with $a + b + c = 0$. A few calculations show that, if $(1, 2).\mathbf{p} = r\mathbf{p}$, and if $(1, 2, 3).\mathbf{p} = s\mathbf{p}$ for some scalars r and s, then $a = b = c = 0$, contradicting $\mathbf{p} \neq 0$.

Exercise 2.33 The set $\{\mathbf{xy}, \mathbf{xz}, \mathbf{yz}\}$ is a convenient basis for $\mathcal{P}_{(1,1,0)}$. It should be clear that the set $\{\mathbf{xy} + \mathbf{xz} + \mathbf{yz}\}$ is a basis for a copy of $\mathcal{I}$ and, say, $\{\mathbf{xy} - \mathbf{xz}, \mathbf{xy} - \mathbf{yz}\}$ is a basis for a copy of $\mathcal{W}$.

Exercise 2.34 Here are two computations to get you started. The vector $(1, 2).\mathbf{w}_1 = -\mathbf{w}_1$ and $(1, 2).\mathbf{w}_2 = \mathbf{w}_2$, so with respect to this basis the matrix realization is

$$\widetilde{(1, 2)} = \begin{pmatrix} -1 & 0 \\ 0 & 1 \end{pmatrix}.$$

Now $(2, 3).\mathbf{w}_1 = \frac{1}{\sqrt{2}}(x - z) = a\mathbf{w}_1 + b\mathbf{w}_2$ for some scalars a and b. We can solve for a and b by equating the coefficients for $\mathbf{x}, \mathbf{y}$ and $\mathbf{z}$ to obtain

$$(1, 2).\mathbf{w}_1 = \frac{1}{2}\mathbf{w}_1 + \frac{\sqrt{3}}{2}\mathbf{w}_2.$$

Similarly, $(2, 3).\mathbf{w}_2 = \frac{\sqrt{3}}{2}\mathbf{w}_1 + \frac{-1}{2}\mathbf{w}_2$, so the matrix realization for this case is

$$\widetilde{(2, 3)} = \begin{pmatrix} \frac{1}{2} & \frac{\sqrt{3}}{2} \\ \frac{\sqrt{3}}{2} & -\frac{1}{2} \end{pmatrix}.$$

Check that these matrices are all unitary, which is the same thing as orthogonal since the matrix entries are real numbers. Compare these results with the results of Exercise 1.90. By the way, note that the vectors $\mathbf{w}_1$ and $\mathbf{w}_2$ are orthonormal with respect to the naive inner product obtained by declaring that the basis vectors $\{\mathbf{x}, \mathbf{y}, \mathbf{z}\}$ are mutually orthonormal.

Intertwining Maps, Complete Reducibility, and Invariant Inner Products

3

We introduce the relevant linear maps that also preserve the algebraic structure of group representations. We use these intertwining maps to prove Maschke's theorem, that every representation of a finite group on a complex vector space is completely reducible. We then review and expand on the notion of inner product spaces, and use G-invariant inner products to give another proof of Maschke's theorem. Finally, we discuss representations on dual spaces.

3.1 Intertwining Maps

Recall from linear algebra that for vector spaces V and W, the set of all linear maps from V to W is itself a vector space, denoted $\mathrm{Hom}(V, W)$, and when V and W are both finite-dimensional, then $\mathrm{Dim}\,\mathrm{Hom}(V, W) = (\mathrm{Dim}\,V)(\mathrm{Dim}\,W)$. Along the same vein as linear maps between vector spaces, and group homomorphisms between groups, we have maps between representations of a group that respect the algebraic structure.

Definition* 3.1 Let (ρ, V) and (π, W) be two representations of a group G. A linear transformation $\phi \colon V \to W$, such that

$$\phi(\rho(g)\mathbf{v}) = \pi(g)\phi(\mathbf{v}) \text{ for all } g \in G, \text{ and } \mathbf{v} \in V,$$

is called an *intertwining map* (also called a *G-intertwining map*, a *G-module homomorphism*, a *G-linear map*, a *G-equivariant map* or just a *G-map*).

If ϕ is also a bijection (and hence invertible), then ϕ is a *G-isomorphism*, and two representations are *isomorphic* or *equivalent* if there is an isomorphism between them. The set of all G-intertwining maps from V to W is denoted $\mathrm{Hom}_G(V, W)$, and is a vector subspace of $\mathrm{Hom}(V, W)$. See Exercise 3.6.

© The Author(s), under exclusive license to Springer Nature Switzerland AG 2022
R. M. Howe, *An Invitation to Representation Theory*, SUMS Readings,
https://doi.org/10.1007/978-3-030-98025-2_3

Notation 3.2 We will use the term *G-map* for simplicity, and we will say that two *G*-isomorphic representations of a group are *equivalent* (denoted $\cong$) to avoid confusion with other types of isomorphisms.

Notation 3.3 Using the simplified "lower dot" notation, we can write $\phi(g.\mathbf{v}) = g.\phi(\mathbf{v})$, being careful to remember that the *G*-action on the left and the *G*-action on the right may be different.

As with linear maps and group homomorphisms, we define the *kernel* of a *G*-map ϕ as $\mathrm{Ker}\,\phi := \{\mathbf{v} \in V \mid \phi(\mathbf{v}) = 0_W\}$, where 0_W is the additive identity element in W. We define the *image* of ϕ as $\mathrm{Im}\,\phi := \{\mathbf{w} \in W \mid \mathbf{w} = \phi(\mathbf{v})$ for some $\mathbf{v} \in V\}$.

Example 3.4 Recall that every linear map is completely determined by its values on the basis elements (and then extended by linearity). Here are some explicit examples of S_3-maps between our polynomial spaces:

- The map $\phi: \mathcal{P}_{(1,0,0)} \to \mathcal{P}_{(2,0,0)}$ defined on the basis elements by

$$\mathbf{x} \mapsto \mathbf{x}^2, \quad \mathbf{y} \mapsto \mathbf{y}^2, \quad \mathbf{z} \mapsto \mathbf{z}^2.$$

- The map $\phi_1: \mathcal{P}_{(1,0,0)} \to \mathcal{P}_{(2,1,0)}$ defined on the basis elements by

$$\mathbf{x} \mapsto \mathbf{xy}^2 + \mathbf{xz}^2, \quad \mathbf{y} \mapsto \mathbf{x}^2\mathbf{y} + \mathbf{yz}^2, \quad \mathbf{z} \mapsto \mathbf{x}^2\mathbf{z} + \mathbf{y}^2\mathbf{z}.$$

- The map $\phi_2: \mathcal{P}_{(1,0,0)} \to \mathcal{P}_{(2,1,0)}$ defined on the basis elements by

$$\mathbf{x} \mapsto \mathbf{x}^2\mathbf{y} + \mathbf{x}^2\mathbf{z}, \quad \mathbf{y} \mapsto \mathbf{xy}^2 + \mathbf{y}^2\mathbf{z}, \quad \mathbf{z} \mapsto \mathbf{xz}^2 + \mathbf{yz}^2.$$

Exercise* 3.5 Verify the above examples. Can you find more S_n-maps between the polynomial subspaces of various types?

Exercise* 3.6 Verify that $\mathrm{Hom}_G(V, W)$ is a vector subspace of $\mathrm{Hom}(V, W)$. Show that if $\phi: V \to W$ is a *G*-map that is invertible, then $\phi^{-1}: W \to V$ is also a *G*-map.

The next few exercises require some calculation. A computer algebra system may be helpful.

Exercise* 3.7 Show that the space $\mathrm{Hom}_{S_3}(\mathcal{P}_{(1,0,0)}, \mathcal{P}_{(2,0,0)})$ is two-dimensional.

Exercise* 3.8 Show that the space $\mathrm{Hom}_{S_3}(\mathcal{P}_{(1,0,0)}, \mathcal{P}_{(2,1,0)})$ is three-dimensional.

Exercise* 3.9 Let $\mathcal{I}$ be the trivial representation in $\mathcal{P}_{(1,0,0)}$ from Example 2.29. Show that the space $\mathrm{Hom}_{S_3}(\mathcal{I}, \mathcal{P}_{(2,1,0)})$ is one dimensional.

Exercise* 3.10 Let $\mathcal{W}$ be the standard representation in $\mathcal{P}_{(1,0,0)}$ from Example 2.29. Show that the space $\mathrm{Hom}_{S_3}(\mathcal{W}, \mathcal{P}_{(2,1,0)})$ is two-dimensional.

Exercise 3.11 Show that a G-map $\phi \colon V \to W$ is injective (one-to-one) if and only if $\mathrm{Ker}\, \phi = \{0\}$. This is a standard result from linear algebra.

Exercise* 3.12 Show that if V is a representation of G, and if U is a G-invariant subspace of V such that $V = U \oplus W$ for some W, then the projection onto U is a G-map. Conclude that the map in Exercise 2.28 is a G-map.

Proposition 3.13 *If $\phi \colon V \to W$ is a G-map, then $\mathrm{Ker}\, \phi$ and $\mathrm{Im}\, \phi$ are G-invariant subspaces of V and W respectively.*

Exercise* 3.14 Prove Proposition 3.13. Hint: Consider the analogous proposition for linear maps between vector spaces, and then apply the definitions involved.

The following is stated as a lemma for historical reasons, but it is essentially a corollary to Proposition 3.13.

Lemma 3.15 (Schur's lemma) *If V and W are irreducible representations of G, and if $\phi \colon V \to W$ is a G-map, then either ϕ is identically zero, i.e., the zero-map, or ϕ is an isomorphism of G-spaces.*

Proof Since $\mathrm{Ker}\, \phi$ is an invariant subspace of V, and since V is irreducible, $\mathrm{Ker}\, \phi$ is either equal to $\{0\}$ or all of V, and hence ϕ is either injective (Exercise 3.11) or the zero-map. If ϕ is non-zero and hence injective, and since $\mathrm{Im}\, \phi$ is an invariant subspace of W which is irreducible, it follows that $\mathrm{Im}\, \phi = W$, and therefore $\phi \colon V \to W$ is a G-isomorphism. $\qquad\square$

The next lemma emphasises the utility of complex scalars.

Lemma 3.16 (Schur's Lemma, Strong Version) *If V is a complex vector space that is an irreducible representation of G, and if $\phi \colon V \to V$ is a non-zero G-map, then ϕ is a scalar multiple of the identity.*

Proof Since ϕ is not identically zero and the complex numbers are algebraically closed, ϕ has a non-zero eigenvalue c with corresponding eigenvector $\mathbf{v}_c$. Let $I \colon V \to V$ be the identity map. Since ϕ is a G-map, the map $\phi - cI$ is also a G-map, with $\mathbf{v}_c \in \mathrm{Ker}(\phi - cI)$. Consequently, $\phi - cI$ cannot be a G-isomorphism by Exercise 3.11, and therefore must be identically zero. That is, $\phi = cI$. $\qquad\square$

Schur's lemma yields the following important corollaries.

Corollary 3.17 *Let V and W be two representations of G, with V irreducible. Then* $\text{Dim Hom}_G(V, W) = 0$ *if and only if W contains no subrepresentation equivalent to V.*

Corollary 3.18 *Let V be a complex vector space that carries a representation of G. Then* $\text{Dim Hom}_G(V, V) = 1$ *if and only if V is irreducible.*

Exercise* 3.19 Prove Corollaries 3.17 and 3.18.

Exercise* 3.20 Use the G-maps from Exercise 3.5 to obtain two distinct irreducible subspaces of $\mathcal{P}_{(2,1,0)}$ equivalent to the standard representation $\mathcal{W}$ in $\mathcal{P}_{(1,0,0)}$.

Here is another collection of interesting intertwining maps.

Exercise 3.21 Define $\mathbb{X}\colon \mathcal{P}(x, y, z) \to \mathcal{P}(x, y, z)$ by $\mathbb{X}(\mathbf{p}) = (\mathbf{x}+\mathbf{y}+\mathbf{z})\mathbf{p}$. In other words, multiply the polynomial $\mathbf{p}$ by $(\mathbf{x} + \mathbf{y} + \mathbf{z})$. Show that $\mathbb{X}$ is an S_3-map. Apply this map to some of the irreducible subspaces previously obtained and describe the results.

Exercise 3.22 Define $\mathbb{D}\colon \mathcal{P}(x, y, z) \to \mathcal{P}(x, y, z)$ by $\mathbb{D}(\mathbf{p}) = (\partial_x + \partial_y + \partial_z)\mathbf{p}$. In other words, take the partial derivatives of $\mathbf{p}$ with respect to each variable, and then add them together. Show that $\mathbb{D}$ is an S_3-map. Apply this map to some of the irreducible subspaces previously obtained and describe the results.

Exercise* 3.23 Define $\mathfrak{F}\colon \mathcal{P}(x, y, z) \to \mathcal{P}(x, y, z)$ by

$$\mathfrak{F}(\mathbf{p}) = (x\partial_y + x\partial_z + y\partial_x + y\partial_z + z\partial_x + z\partial_y)\mathbf{p}.$$

For example, the operation $x\partial_y(\mathbf{p})$ means "take the partial derivative of $\mathbf{p}$ with respect to y, then multiply the result by x." Show that $\mathfrak{F}$ is an S_3-map. Apply this map to some of the irreducible subspaces previously obtained and describe the results. A computer algebra system may be useful here.

3.2   Complete Reducibility

We saw in Example 2.29 that $\mathcal{P}_{(1,0,0)} = \mathcal{I} \oplus \mathcal{W}$ decomposes into the direct sum of two irreducible subspaces. A representation that is a direct sum of irreducible representations is said to be *completely reducible*.

Definition 3.24 If V is a representation of G and if U is a G-invariant subspace of V, then a *complementary subspace* is a G-invariant subspace W such that $V = U \oplus W$.

In Example 2.29, W and $\mathcal{I}$ are complementary subspaces in $\mathcal{P}_{(1,0,0)}$.

Theorem 3.25 (Maschke's Theorem)   *If V is a complex[1] vector space that carries a representation of a finite group G, and if U is a G-invariant subspace of V, then there is a complementary subspace W of V such that $V = U \oplus W$.*

Proof   This proof skips some details, but you should work through them and verify the assertions.

The idea is to construct a projection of V onto U that is also a G-map. Using Exercise 2.27 and Proposition 3.13, it follows that the kernel W will be a complementary G-invariant subspace with $V = U \oplus W$.

Let $P_0: V \rightarrow U$ be any projection of V onto U. We then construct a new projection P (averaging P_0 over G) by defining

$$P(\mathbf{v}) := \frac{1}{|G|} \sum_{g \in G} g.[P_0(g^{-1}.\mathbf{v})].$$

Then P is the desired G-map from V whose image is U, and whose kernel W is the desired complementary subspace.   □

Corollary 3.26   *Every finite-dimensional complex representation of a finite group is completely reducible.*

Exercise* 3.27   Prove Corollary 3.26.

Non-Example 3.28   Consider the representation of the group $(\mathbb{R}, +)$ on $\mathbb{R}^2$ given by

$$t.(x, y) = (x + ty, y) = \begin{pmatrix} 1 & t \\ 0 & 1 \end{pmatrix} \begin{bmatrix} x \\ y \end{bmatrix}.$$

Then the x-axis is an invariant subspace with no complementary subspace.

Exercise 3.29   Verify the claims made in Non-Example 3.28: that the representation is, in fact, a representation of $(\mathbb{R}, +)$, and that the x-axis is an invariant subspace with no complementary subspace. Also show that the projection onto the x-axis is not a G-map. Hint: Any complementary subspace must be one-dimensional, and thus spanned by a vector $(a, b) \in \mathbb{R}^2$. But this subspace is not G-invariant unless $b = 0$. Why does this not contradict Maschke's Theorem?

Remark* 3.30   Since each linear map between finite-dimensional vector spaces can be realized as matrix multiplication of coordinate (column) vectors (Defini-

---

[1] To be consistent, we use $\mathbb{C}$ as the field of scalars. The necessary property is that $|G| \neq 0$, as is the case when the field of scalars has characteristic 0.

tion 1.87), Maschke's Theorem can also be stated in terms of matrices. In this context, the theorem says that if V is a representation of a finite group G, then there is a basis for V for which the matrix realization of this group action can be written in block-diagonal form, where each block is the matrix realization of an irreducible representation.

For example, consider the permutation representation of S_3 on $\mathcal{P}_{(1,0,0)} = \mathcal{I} \oplus \mathcal{W}$. Using $\mathbf{w_1} = \mathbf{x} + \mathbf{y} + \mathbf{z}$ as a basis for $\mathcal{I}$, and using $\mathbf{w_2} = \mathbf{x} - \mathbf{y}$ and $\mathbf{w_3} = \mathbf{x} - \mathbf{z}$ as a basis for $\mathcal{W}$, we obtain a basis $\{\mathbf{w_1}, \mathbf{w_2}, \mathbf{w_3}\}$ for $\mathcal{P}_{(1,0,0)}$. We see that, for example $(1,2).\mathbf{w_1} = \mathbf{w_1}$, $(1,2).\mathbf{w_2} = -\mathbf{w_2}$ and $(1,2).\mathbf{w_3} = -\mathbf{w_2} + \mathbf{w_3}$. Thus, with respect to this basis, we obtain the matrix realization

$$\widetilde{(1,2)} = \begin{pmatrix} 1 & 0 & 0 \\ 0 & -1 & -1 \\ 0 & 0 & 1 \end{pmatrix}.$$

You should try this with other elements of S_3 and/or with another basis for $\mathcal{W}$.

3.3   Invariant Inner Products and Another Proof of Complete Reducibility

Recall that an *inner product* (or a *scalar product*) on a real vector space V is an operation that combines two vectors $\mathbf{v}, \mathbf{w}$ in V to produce a real number, denoted $\langle \mathbf{v}, \mathbf{w} \rangle$. This product satisfies the following properties for all $\mathbf{v}, \mathbf{w}, \mathbf{x} \in V$ and any $r, s \in \mathbb{R}$:

(1) $\langle \mathbf{v}, \mathbf{v} \rangle \geq 0$ Positive-definite.
(2) $\langle \mathbf{v}, \mathbf{w} \rangle = \langle \mathbf{w}, \mathbf{v} \rangle$ Symmetric.
(3) $\langle r\mathbf{v} + s\mathbf{w}, \mathbf{x} \rangle = r\langle \mathbf{v}, \mathbf{x} \rangle + s\langle \mathbf{w}, \mathbf{x} \rangle$ Bilinear.

While property (3) only guarantees linearity in the first position, together with property (2) we have bilinearity; that is, $\langle \mathbf{v}, r\mathbf{w} + s\mathbf{x} \rangle = r\langle \mathbf{v}, \mathbf{w} \rangle + s\langle \mathbf{v}, \mathbf{x} \rangle$ also holds. We often see the term *positive-definite symmetric bilinear form* for this type of inner product. An inner product is said to be *non-degenerate* if, in addition, it also satisfies the condition

(4) If $\mathbf{v} \in V$, and if $\langle \mathbf{v}, \mathbf{w} \rangle = \mathbf{0}$ for all $\mathbf{w} \in V$, then $\mathbf{v} = \mathbf{0}$.

For complex vector spaces, we replace symmetry (2) with *conjugate symmetry*,

$$\langle \mathbf{v}, \mathbf{w} \rangle = \overline{\langle \mathbf{w}, \mathbf{v} \rangle},$$

and the inner product is said to be *Hermitian*. As a result, if a Hermitian form is linear in the first position, then it is conjugate-linear in the second position, that is, $\langle \mathbf{v}, s\mathbf{w} \rangle = \bar{s} \langle \mathbf{v}, \mathbf{w} \rangle$, and the label bilinear is replaced by *sesquilinear*.

A vector space with an inner product is called an *inner product space*. The quantity $\|\mathbf{v}\| = \sqrt{\langle \mathbf{v}, \mathbf{v} \rangle}$ is called the *norm* of the vector $\mathbf{v}$, and the properties of an inner product guarantee that $\langle \mathbf{v}, \mathbf{v} \rangle$, and hence $\|\mathbf{v}\|$, is a non-negative real number. Two vectors $\mathbf{v}$ and $\mathbf{w}$ are said to be *orthogonal* if $\langle \mathbf{v}, \mathbf{w} \rangle = 0$, and *orthonormal* if $\langle \mathbf{v}, \mathbf{w} \rangle = \delta_{\mathbf{v}, \mathbf{w}}$ (Kronecker delta, see Remark 1.93). We know from linear algebra that inner products generalize the geometric notions of length and angle.

Exercise* 3.31 Verify that if a Hermitian inner product is linear in the first position, then it is conjugate-linear in the second position. Look up the meaning of the prefix "sesqui."

Example* 3.32 We list some examples of inner product spaces at the end of the chapter.

An inner product is not needed to define a norm on a vector space.

Remark* 3.33 Given a real or complex vector space V, a *norm* on V is a real-valued function $\mathbf{v} \mapsto \|\mathbf{v}\|$ that has the following properties.

(1) For every $\mathbf{v} \in V$, $\|\mathbf{v}\| \geq 0$.
(2) $\|\mathbf{v}\| = 0$ if and only if $\mathbf{v} = 0$.
(3) For every $\mathbf{v} \in V$ and every scalar a, $\|a\mathbf{v}\| = |a| \|\mathbf{v}\|$.
(4) The *triangle equality* holds; that is, for every $\mathbf{v}, \mathbf{w} \in V$, we have

$$\|\mathbf{v} + \mathbf{w}\| \leq \|\mathbf{v}\| + \|\mathbf{w}\|.$$

By the above paragraphs, any inner-product space can be a normed space, but there are normed vector spaces that are not inner-product spaces.

Exercise 3.34 You should verify that (at least some of) the examples in Example 3.32 and Remark 3.33 satisfy the properties of inner products and norms, respectively.

Exercise 3.35 Show that the rows (columns) of an $n \times n$ orthogonal matrix are orthonormal with respect to the standard dot product on $\mathbb{R}^n$. Show that the rows (columns) of an $n \times n$ unitary matrix are orthonormal with respect to the standard Hermitian product on $\mathbb{C}^n$.

Certain inner products are useful in representation theory.

Definition 3.36 Let V be an inner product space that is a representation of a group G. Then the inner product is a *G-invariant inner product* if $\langle g.\mathbf{v}, g.\mathbf{w} \rangle = \langle \mathbf{v}, \mathbf{w} \rangle$ for all $\mathbf{v}, \mathbf{w} \in V$, and for all $g \in G$.

Exercise 3.37 Let G be the orthogonal group $O(n, \mathbb{R})$ of Exercise 1.57. Show that the usual dot product on $\mathbb{R}^n$ is G-invariant. Hint: Write $\mathbf{v}$ and $\mathbf{w}$ as column vectors so that $\langle \mathbf{v}, \mathbf{w} \rangle = \mathbf{v}^T \mathbf{w}$, where $\mathbf{v}^T$ denotes the matrix transpose of $\mathbf{v}$, and where the operation on the right-hand side is matrix multiplication.

Definition 3.38 We can construct an inner product on any finite-dimensional real or complex vector space V by choosing a basis for V and declaring that the basis vectors are orthonormal, in which case this inner product just reduces to the dot or Hermitian inner product on the coordinate vectors. For lack of a better term, let's call this the *naive inner product* on V. For the polynomial spaces that we are working with, we have the obvious monomial bases which provides us with a naive inner product.

Exercise* 3.39 Let S_n act on $\mathcal{P}(x_1, \ldots, x_n)$ as usual. Show that the naive inner product defined above is S_n-invariant.

Exercise* 3.40 Let V be a representation of a group G with a G-invariant inner product. Suppose W is a subspace of V, and let $W^\perp = \{\mathbf{v} \in V \mid \langle \mathbf{v}, \mathbf{w} \rangle = 0 \text{ for all } \mathbf{w} \in W\}$ denote the orthogonal complement of W in V. Show that if W is a G-invariant subspace of V, then so is $W^\perp$. Show that if V is finite-dimensional, then $V = W \oplus W^\perp$. Hint: You may need to review some linear algebra.

Exercise 3.41 Let $\mathcal{P}_1(V)$ be the representation of S_3 given in Example 2.29 along with the naive inner product. Check that $\mathcal{I}^\perp = \mathcal{W}$, and that $\mathcal{W}^\perp = \mathcal{I}$.

Inner products, along with Exercise 3.40, provide us with another proof of Maschke's Theorem. If a representation V of a group G has a G-invariant inner product, then any invariant subspace W of V will have a complementary subspace, namely $W^\perp$ by Exercise 3.40.

Not all inner products are G-invariant, but given any inner product $\langle\,,\,\rangle$ on V we can construct a G-invariant inner product $\langle\,,\,\rangle'$ by averaging over the finite group G:

Equation 3.42

$$\langle \mathbf{v}, \mathbf{w} \rangle' := \sum_{g \in G} \langle g.\mathbf{v}, g.\mathbf{w} \rangle.$$

Exercise* 3.43 Show that $\langle\,,\,\rangle'$ is in fact an inner product that is G-invariant.

Remark 3.44 An inner product space that is "complete" is called a *Hilbert space*, after the German mathematician David Hilbert. Any finite-dimensional inner product space is complete, and is therefore a Hilbert space. In Example 3.32, $L^2([-\pi, \pi])$ and Fock space are important examples of infinite-dimensional Hilbert spaces. There are numerous applications of Hilbert spaces, including ordinary and partial differential equations, Fourier analysis, ergodic theory, probability, and quantum mechanics.

A representation of a group G on a Hilbert space $\mathcal{H}$ for which the inner product is G-invariant is called a *unitary* representation, and thus every invariant subspace has a complementary subspace.[2] The group G need not be finite and $\mathcal{H}$ need not be finite dimensional, but there are versions of "Maschke's Theorem" for unitary representations.

The above discussion, along with Eq. 3.42, essentially says that any representation of a finite group on an inner-product space is unitarizable, and thus the matrix realization can be by unitary matrices (Exercise 1.61). Unitary representations have important applications in quantum physics, harmonic analysis, and geometric and algebraic topology.

Exercise* 3.45 This exercise strays into the realm of analysis. Consider the inner product space $C[0, 1]$ from Example 3.32. Let $f_n(x) := x^n$ and define

$$f(x) := \begin{cases} 0 & \text{if } x \in [0, 1); \\ 1 & \text{if } x = 1. \end{cases}$$

Show that the sequence of functions $\{f_n\}$ converges to f in the norm defined by the inner product on $C[0, 1]$. Since f is not continuous, that is $f \notin C[0, 1]$, conclude that $C[0, 1]$ is not complete, and thus is not a Hilbert space.

3.4   Dual Spaces and Contragredient Representations

We wish to expand on the discussion from Remark 1.93 regarding dual spaces. For a complex vector space V with basis $\{v_1, \dots v_m\}$, we define V^*, the *dual space* of V as,

$$V^* := \{\phi \colon V \to \mathbb{C} \mid \phi \text{ is linear}\}.$$

Vectors in V^* are called *linear functionals*.

---

[2] Subject to certain topological considerations.

Exercise 3.46 Let V be a Hermitian inner product space, and fix a vector $\mathbf{v}_0 \in V$. Confirm that the map $\phi_{\mathbf{v}_0}$ defined by $\phi_{\mathbf{v}_0}(\mathbf{v}) := \langle \mathbf{v}, \mathbf{v}_0 \rangle$ is in V^*, but the map $\mathbf{v} \mapsto \langle \mathbf{v}_0, \mathbf{v} \rangle$ is not in V^*.

By the discussion of Definition 1.87, for any $\phi \in V^*$, and any $\mathbf{v} = r_1 \mathbf{v}_1 + \cdots + r_m \mathbf{v}_m$, there is a $1 \times m$ matrix $\tilde{\phi}$, *i.e*, a row vector, such that

$$\phi(\mathbf{v}) = \begin{bmatrix} a_1, \ldots, a_m \end{bmatrix} \begin{bmatrix} r_1 \\ r_2 \\ \vdots \\ r_m \end{bmatrix},$$

and where $a_i = \phi(\mathbf{v}_i)$. In other words, if we realize V as a space of column coordinate vectors, then we can realize V^* as a space of row coordinate vectors.

Once we recognize the above matrix product as an inner product of coordinate vectors, we have a converse to Exercise 3.46, which roughly says that any $\phi \in V^*$ can be realized as $\phi_{\mathbf{v}_0}(\mathbf{v}) := \langle \mathbf{v}, \mathbf{v}_0 \rangle$ for some $\mathbf{v}_0 \in V$. This is a simple example of the *Riesz representation theorem*.

Exercise* 3.47 Find other examples of linear functionals.

Now let $\{\mathbf{v}_1, \ldots \mathbf{v}_m\}$ and $\{\mathbf{w}_1, \ldots \mathbf{w}_n\}$ V be bases for V and W respectively. For any linear map $F \colon V \to W$, we can define $F^* \colon W^* \to V^*$, called the *dual of F*, by

$$[F^*(\phi)](\mathbf{v}) := \phi \circ F(\mathbf{v}).$$

Sometimes F^* is referred to as the *pullback of ϕ along F* or the *transpose* of F. Translating this into the language of matrix multiplication we have;

$$[F^*(\phi)](\mathbf{v}) = \begin{bmatrix} b_1, \ldots, b_n \end{bmatrix} \begin{bmatrix} & & \\ & \tilde{F} & \\ & & \end{bmatrix} \begin{bmatrix} r_1 \\ r_2 \\ \vdots \\ r_m \end{bmatrix},$$

(as above, $b_i = \phi(\mathbf{w}_i)$, and $\left[b_1, \ldots, b_n\right] = \tilde{\phi}$), so we can think of F^* acting on functionals in W^* as right multiplication on row coordinate vectors. If we want to represent W^* as column coordinate vectors, we take the transpose of the whole mess;

$$[F^*(\phi)](\mathbf{v}) = \left[r_1, \ldots, r_m\right] \left[\quad \tilde{F}^T \quad\right] \begin{bmatrix} b_1 \\ b_2 \\ \vdots \\ b_n \end{bmatrix},$$

so F^* acts by left multiplication on column coordinate vectors via $\tilde{F}^T$.

Given a representation (ρ, V) of G, it is natural to ask how to define a representation ρ^* of G on V^*. Consistency requires that

$$[\rho^*(g)\phi]\rho(g)\mathbf{v} = \phi(\mathbf{v}),$$

and we see that if we define $\rho^*(g) := \rho(g^{-1})^T$, then

$$[\rho^*(g)\phi]\rho(g)\mathbf{v} = [\rho(g^{-1})^T\phi]\rho(g)\mathbf{v} = \phi^T\rho(g^{-1})\rho(g)\mathbf{v} = \phi(\mathbf{v}),$$

as desired. The representation (ρ^*, V^*) is called the *contragredient* or *dual* representation.

Exercise 3.48 Verify that (ρ^*, V^*) is, in fact, a representation of G.

3.5 Hints and Additional Comments

Definition 3.1 Another way to depict the situation where $\phi(\rho(g)\mathbf{v}) = \pi(g)\phi(\mathbf{v})$ is by using what is called a *commutative diagram*,

$$\begin{array}{ccc} V & \xrightarrow{\phi} & W \\ \downarrow{\scriptstyle\rho(g)} & & \downarrow{\scriptstyle\pi(g)} \\ V & \xrightarrow{\phi} & W. \end{array}$$

The diagram is interpreted to mean "we can travel either path from the upper left corner to the lower right corner with the same result." An example that we are all familiar with is the composition of functions, $f \circ g = h$, which we can depict with the diagram

$$
\begin{array}{ccc}
 & A & \\
{\scriptstyle g}\nearrow & & \searrow{\scriptstyle f} \\
B & \xrightarrow{h} & C \,.
\end{array}
$$

Some commutative diagrams can be quite complicated.

Exercise 3.5 The G-maps in the examples can be succinctly described as "x goes to the sum of all monomials in which x appears to the power one", or "x goes to the sum of all monomials in which x appears to the power two", etc. Is this all of them? First you need to decide what is meant by "all."

Exercise 3.6 It is a standard result from linear algebra that $\mathrm{Hom}(V, W)$, the set of all linear maps between two vector spaces V and W, is itself vector space. To show that $\mathrm{Hom}_G(V, W)$ is a subspace of $\mathrm{Hom}(V, W)$ is a routine application of the "Test for Subspaces" and the definitions involved.

To show that the inverse of a G-map is also a G-map, suppose $\phi\colon V \to W$ is a linear isomorphism between vector spaces. It is another standard result from linear algebra that $\phi^{-1}\colon W \to V$ is also linear, and therefore must be a linear isomorphism. Thus it remains to show that if ϕ is a G-map, then ϕ^{-1} is also a G-map. Now certainly

$$
\phi[\phi^{-1}(g.\mathbf{v})] = g.\mathbf{v},
$$

and by hypothesis ϕ is a G-map, so

$$
\phi[g.\phi^{-1}(\mathbf{v})] = g.\phi[(\phi^{-1}(\mathbf{v})] = g.\mathbf{v}.
$$

Equating these two expressions for $g.\mathbf{v}$ we have

$$
\phi[\phi^{-1}(g.\mathbf{v})] = \phi[g.\phi^{-1}(\mathbf{v})].
$$

Recall that if a map ϕ is one-to-one, then $\phi(a) = \phi(b)$ implies that $a = b$. It follows that $\phi^{-1}(g.\mathbf{v}) = g.\phi^{-1}(\mathbf{v})$, so ϕ^{-1} is a G-map.

This result provides the answer to the question posed at the end of the hint for Exercise 3.5. The set of G-maps between any two representations is itself a vector space. Finding "all" such maps boils down to finding a basis for this space.

Exercise 3.7 To show that $\mathrm{Hom}_{S_3}(\mathcal{P}_{(1,0,0)}, \mathcal{P}_{(2,0,0)})$ is two-dimensional, we will find a basis for this space. We will work this out in considerable detail as a model for the other similar exercises.

Let $\psi_0 : \mathcal{P}_{(1,0,0)} \to \mathcal{P}_{(2,0,0)}$ be the map defined on the basis elements $\{\mathbf{x}, \mathbf{y}, \mathbf{z}\}$ by

$$\mathbf{x} \mapsto \mathbf{y}^2 + \mathbf{z}^2, \quad \mathbf{y} \mapsto \mathbf{x}^2 + \mathbf{z}^2 \text{ and } \mathbf{z} \mapsto \mathbf{x}^2 + \mathbf{y}^2,$$

then extend by linearity. As with Exercise 3.5, we can succinctly describe this map as "$\mathbf{x}$ goes to the sum of all monomials in $\mathcal{P}_{(2,0,0)}$ in which $\mathbf{x}$ appears to the power zero", etc. Similarly let $\psi_2 : \mathcal{P}_{(1,0,0)} \to \mathcal{P}_{(2,0,0)}$ be the map defined on the basis elements as

$$\mathbf{x} \mapsto \mathbf{x}^2, \quad \mathbf{y} \mapsto \mathbf{y}^2 \text{ and } \mathbf{z} \mapsto \mathbf{z}^2.$$

It is straightforward to show that these maps are linearly independent: if $a\psi_0 + b\psi_2 = 0$ for scalars a and b then $a\psi_0(\mathbf{v}) + b\psi_2(\mathbf{v}) = 0$ for all $\mathbf{v} \in \mathcal{P}_{(1,0,0)}$. Since these maps are linear by construction, it is sufficient to verify this on the basis vectors $\mathbf{x}$ $\mathbf{y}$, and $\mathbf{z}$.

Now

$$a\psi_0(\mathbf{x}) + b\psi_2(\mathbf{x}) = 0,$$
$$\Rightarrow a(\mathbf{y}^2 + \mathbf{z}^2) + b(\mathbf{x}^2) = 0,$$
$$\Rightarrow \quad b\mathbf{x}^2 + a\mathbf{y}^2 + a\mathbf{z}^2 = 0.$$

Since the set $\{\mathbf{x}^2, \mathbf{y}^2, \mathbf{z}^2\}$ is a basis for $\mathcal{P}_{(2,0,0)}$, we must have $a = b = 0$ (so it is redundant to verify this for $\mathbf{y}$ or $\mathbf{z}$).

To show that these two maps span $\mathrm{Hom}_{S_3}(\mathcal{P}_{(1,0,0)}, \mathcal{P}_{(2,0,0)})$, we use the fact that they are G-maps. Suppose $\psi : \mathcal{P}_{(1,0,0)} \to \mathcal{P}_{(2,0,0)}$ is any G-map. Then, for some scalars $a, b, c, \alpha, \beta, \gamma$;

$$\psi(\mathbf{x}) = a\mathbf{x}^2 + b\mathbf{y}^2 + c\mathbf{z}^2,$$
$$\text{and } \psi(\mathbf{y}) = \alpha\mathbf{x}^2 + \beta\mathbf{y}^2 + \gamma\mathbf{z}^2.$$

Since ψ is a G-map, we must have for, say, the transposition $(1, 2) \in S_3$;

$$\psi((1, 2).\mathbf{x}) = (1, 2).\psi(\mathbf{x}),$$
$$\Rightarrow \qquad \psi(\mathbf{y}) = (1, 2).(a\mathbf{x}^2 + b\mathbf{y}^2 + c\mathbf{z}^2),$$
$$\Rightarrow \alpha\mathbf{x}^2 + \beta\mathbf{y}^2 + \gamma\mathbf{z}^2 = b\mathbf{x}^2 + a\mathbf{y}^2 + c\mathbf{z}^2.$$

Hence $\alpha = b, \beta = a$ and $\gamma = c$, and so $\psi(\mathbf{y}) = b\mathbf{x}^2 + a\mathbf{y}^2 + c\mathbf{z}^2$.
Similarly, for $(1, 2, 3) \in S_3$;

$$\psi((1, 2, 3).\mathbf{x}) = (1, 2, 3).\psi(\mathbf{x}),$$
$$\Rightarrow \quad \psi((1, 2, 3).\mathbf{x}) = (1, 2, 3).(a\mathbf{x}^2 + b\mathbf{y}^2 + c\mathbf{z}^2),$$
$$\Rightarrow \qquad \psi(\mathbf{y}) = b\mathbf{x}^2 + a\mathbf{y}^2 + c\mathbf{z}^2,$$
$$\Rightarrow b\mathbf{x}^2 + a\mathbf{y}^2 + c\mathbf{z}^2 = c\mathbf{x}^2 + a\mathbf{y}^2 + c\mathbf{x}^2.$$

Hence $b = c$.

From the two computations above, we conclude that

$$\psi(\mathbf{y}) = b\mathbf{x}^2 + a\mathbf{y}^2 + b\mathbf{z}^2 = a\psi_0(\mathbf{y}) + b\psi_2(\mathbf{y}),$$

and therefore $\{\psi_0, \psi_2\}$ is a basis for $\mathrm{Hom}_{s_3}(\mathcal{P}_{(1,0,0)}, \mathcal{P}_{(2,0,0)})$. The computations on the other basis elements yield identical results.

Exercise 3.8 Similar to Exercise 3.7, let

$$\begin{aligned}
\psi_0(\mathbf{x}) &= \mathbf{y}^2\mathbf{z} + \mathbf{y}\mathbf{z}^2, \quad \text{etc.,} \\
\psi_1(\mathbf{x}) &= \mathbf{x}\mathbf{y}^2 + \mathbf{x}\mathbf{z}^2, \quad \text{etc.,} \\
\psi_2(\mathbf{x}) &= \mathbf{x}^2\mathbf{y} + \mathbf{x}^2\mathbf{z}, \quad \text{etc.}
\end{aligned}$$

Then $\{\psi_0, \psi_1, \psi_2\}$ is a basis for $\mathrm{Hom}_{s_3}(\mathcal{P}_{(1,0,0)}, \mathcal{P}_{(2,1,0)})$. The proof is essentially the same as Exercise 3.7, but the computations are somewhat more involved.

Exercise 3.9 Let ψ_0, ψ_1 and ψ_2 be as in Exercise 3.8. Then any G-map from $\mathcal{P}_{(1,0,0)}$ to $\mathcal{P}_{(2,1,0)}$, and hence any G-map from $\mathcal{I}$ to $\mathcal{P}_{(2,1,0)}$, can be written as a linear combination of ψ_0, ψ_1 and ψ_2. Now check that $\psi_0(\mathbf{x} + \mathbf{y} + \mathbf{z}) = \psi_1(\mathbf{x} + \mathbf{y} + \mathbf{z}) = \psi_2(\mathbf{x} + \mathbf{y} + \mathbf{z})$ so that any one of the maps ψ_0, ψ_1 or ψ_2 can serve as a basis for $\mathrm{Hom}_{s_3}(\mathcal{I}, \mathcal{P}_{(2,1,0)})$, which is consequently one-dimensional.

Exercise 3.10 Again, let ψ_0, ψ_1 and ψ_2 be as in Exercise 3.8. Then any G-map from $\mathcal{W}$ to $\mathcal{P}_{(2,1,0)}$ can be written as a linear combination of ψ_0, ψ_1 and ψ_2. Choose a basis, say, $\{(\mathbf{x} - \mathbf{y}), (\mathbf{x} - \mathbf{z})\}$, for $\mathcal{W} \subset \mathcal{P}_{(1,0,0)}$. Now observe that the vectors

$$\psi_0(\mathbf{x} - \mathbf{y}), \quad \psi_1(\mathbf{x} - \mathbf{y}), \quad \text{and} \quad \psi_2(\mathbf{x} - \mathbf{y})$$

are pairwise linearly independent, but that

$$\psi_0(\mathbf{x} - \mathbf{y}) + \psi_1(\mathbf{x} - \mathbf{y}) + \psi_2(\mathbf{x} - \mathbf{y}) = 0.$$

The same relations hold when applied to the other basis vector $(\mathbf{x} - \mathbf{z})$. Thus $\psi_0 + \psi_1 + \psi_2 = 0$, and so any one of these maps is a linear combination of the other two, and hence $\mathrm{Hom}_{s_3}(\mathcal{W}, \mathcal{P}_{(2,1,0)})$ is 2-dimensional.

Exercise 3.12 To show that $g.[P_U(\mathbf{v})] = P_U(g.\mathbf{v})$, we compute

$$\begin{aligned}
g.[P_U(\mathbf{v})] &= g.[P_U(\mathbf{u} + \mathbf{w})] \quad \text{(since } \mathbf{v} = \mathbf{u} + \mathbf{w}, \text{ for some } \mathbf{u} \in U, \mathbf{w} \in W) \\
&= g.\mathbf{u}. \qquad\qquad\quad \text{(since } P_U(\mathbf{u} + \mathbf{w}) = \mathbf{u})
\end{aligned}$$

On the other hand,

$$P_U(g.\mathbf{v}) = P_U(g.[\mathbf{u} + \mathbf{w}])$$
$$= P_U(g.\mathbf{u} + g.\mathbf{w}) \text{ (since } g \text{ acts linearly)}$$
$$= g.\mathbf{u}. \qquad \text{(since } U \text{ and } W \text{ are } G\text{-invariant)}$$

Exercise 3.14 If $\phi: V \to W$ is a G-map, we know from linear algebra that Ker ϕ is a subspace of V. To show that Ker ϕ is G-invariant, let $\mathbf{v} \in$ Ker ϕ. Then $\phi(g.\mathbf{v}) = g.\phi(\mathbf{v}) = g.\mathbf{0} = \mathbf{0}$. Thus $g.\mathbf{v} \in$ Ker ϕ, and so Ker ϕ is G-invariant.

Exercise 3.19 Corollary 3.17: Suppose that V and W are two representations of G, with V irreducible. If Dim Hom$_G(V, W) \neq 0$, then there is a non-zero G-map $\phi: V \to W$. Since V is irreducible, $\phi(V) \cong V$ by Schur's lemma, and so $\phi(V)$ is a subrepresentation of W equivalent to V. Contrapositively, if W contains no subrepresentation equivalent to V, then Dim Hom$_G(V, W) = 0$.

Conversely, suppose that W contains a subrepresentation equivalent to V. That is, $W \cong V' \oplus U$, with $V' \cong V$. Then there is a G-isomorphism $\phi: V \to V' \subset W$, and so Dim Hom$_G(V, W) \geq 1$. Contrapositively, if Dim Hom$_G(V, W) = 0$, then W contains no subrepresentation equivalent to V.

Corollary 3.18: If V_1 and V_2 are subrepresentations, and if $V \cong V_1 \oplus V_2$, then the projection maps $P_{V_1}: V \to V_1 \subset V$ and $P_{V_2}: V \to V_2 \subset V$ are G-maps by Exercise 3.12 that are linearly independent (check this), and hence Dim Hom$_G(V, V) \geq 2$.

Conversely, if V is irreducible, then any $\phi \in$ Hom$_G(V, V)$ is a scalar multiple of the identity map by the strong version of Schur's lemma. Therefore Hom$_G(V, V)$ is spanned by the identity, and so Dim Hom$_G(V, V) = 1$. Note that this is the only case requiring complex scalars.

Exercise 3.20] If we use the basis $\{\mathbf{x} - \mathbf{y}, \mathbf{x} - \mathbf{z}\}$ for $\mathcal{W} \subseteq \mathcal{P}_1$ then

$$\mathbf{w}_1 = \phi_1(\mathbf{x} - \mathbf{y}) = xy^2 + xz^2 - x^2y - yz^2,$$

and

$$\mathbf{w}_2 = \phi_1(\mathbf{x} - \mathbf{z}) = xy^2 + xz^2 - x^2z - y^2z.$$

Since these two vectors are linearly independent, they must span an irreducible subspace equivalent to $\mathcal{W}$ by Schur's lemma. Now

$$\mathbf{w}_3 = \phi_2(\mathbf{x} - \mathbf{y}) = x^2y + x^2z - xy^2 - y^2z,$$

and

$$\mathbf{w}_4 = \phi_2(\mathbf{x} - \mathbf{z}) = x^2y + x^2z - xz^2 - yz^2$$

are also linearly independent, and so must also span a copy of $\mathcal{W}$. Check that the set $\{\mathbf{w}_1, \mathbf{w}_2, \mathbf{w}_3, \mathbf{w}_4\}$ is linearly independent in $\mathcal{P}_{(2,1,0)}$, and therefore $\{\mathbf{w}_1, \mathbf{w}_2\}$ and $\{\mathbf{w}_3, \mathbf{w}_4\}$ must be bases for two separate copies of $\mathcal{W}$. By the way, check that $\phi_1(\mathbf{x} + \mathbf{y} + \mathbf{z}) = \phi_2(\mathbf{x} + \mathbf{y} + \mathbf{z})$. Why?

Exercise 3.23 First notice that, say, $x\partial_y(x^2) = 0$ and that $x\partial_y(y^2) = 2xy$. Some calculation shows that $\mathfrak{F}(x^2 + y^2 + z^2) = 4xy + 4xz + 4yz$, so $\mathfrak{F}$ maps the trivial representation in $\mathcal{P}_{(2,0,0)}$ to the trivial representation in $\mathcal{P}_{(1,1,0)}$.

Applying $\mathfrak{F}$ to the vectors $\mathbf{w}_1$ and $\mathbf{w}_2$ from Exercise 3.20 yields

$$\mathfrak{F}(\mathbf{w}_1) = \mathbf{y}^3 - \mathbf{x}^3 + 2\mathbf{x}^2\mathbf{y} - 2\mathbf{x}\mathbf{y}^2 + \mathbf{x}^2\mathbf{z} - \mathbf{x}\mathbf{z}^2 + \mathbf{y}\mathbf{z}^2 - \mathbf{y}^2\mathbf{z},$$

and

$$\mathfrak{F}(\mathbf{w}_2) = \mathbf{z}^3 - \mathbf{x}^3 + 2\mathbf{x}^2\mathbf{z} - 2\mathbf{x}\mathbf{z}^2 + \mathbf{x}^2\mathbf{y} - \mathbf{x}\mathbf{y}^2 + \mathbf{y}^2\mathbf{z} - \mathbf{y}\mathbf{z}^2.$$

By Schur's lemma $\mathfrak{F}(\mathbf{w}_1)$ and $\mathfrak{F}(\mathbf{w}_2)$ span a copy of the irreducible representation $\mathcal{W}$ in $\mathcal{P}_{(3,0,0)} \oplus \mathcal{P}_{(2,1,0)}$.

Can you generalize the maps $\mathbb{X}$, $\mathbb{D}$ and $\mathfrak{F}$?

Exercise 3.27 The proof should be clear for more mathematically experienced readers, but we present it for the benefit of the novice mathematicians.

Let V be a finite-dimensional representation of a finite group G. If V is irreducible, then we are done. If not, then by Maschke's theorem, V has an irreducible subspace U with complementary subspace W. If W is irreducible, then we are done. If not, we apply Maschke's theorem to W, and continue by induction. Since V is finite dimensional, this process has to end, so that V is a direct sum of irreducible subrepresentations.

Remark 3.30 A block diagonal matrix is a square matrix, with main diagonal blocks that are square matrices, and the off-diagonal blocks are zero matrices. More explicitly, a block diagonal matrix has the form

$$\begin{pmatrix} \mathbf{A}_1 & 0 & \cdots & 0 \\ 0 & \mathbf{A}_2 & \cdots & 0 \\ \vdots & \vdots & \ddots & \vdots \\ 0 & 0 & \cdots & \mathbf{A}_k \end{pmatrix},$$

where each $\mathbf{A}_i$ is a square matrix. Matrix computations such as products and inverses, as well as theoretical considerations, are more transparent with block diagonal matrices.

Exercise 3.31 If $\langle r\mathbf{v}, \mathbf{w} \rangle = r\langle \mathbf{v}, \mathbf{w} \rangle$, and if $\langle \mathbf{v}, \mathbf{w} \rangle = \overline{\langle \mathbf{w}, \mathbf{v} \rangle}$, then

$$\langle \mathbf{v}, s\mathbf{w} \rangle = \overline{\langle s\mathbf{w}, \mathbf{v} \rangle} = \overline{s}\,\overline{\langle \mathbf{w}, \mathbf{v} \rangle} = \overline{s}\langle \mathbf{v}, \mathbf{w} \rangle.$$

It in usual in the physics literature for a Hermitian inner product be linear in the second position, in which case it is conjugate-linear in the first position. This inner product is typically denoted using *Dirac bracket notation*: $\langle \mathbf{v} \mid \mathbf{w} \rangle$. The vector $\langle \mathbf{v} \mid$ is referred to as a "bra" vector in V^*, while $\mid \mathbf{w} \rangle$ is called a "ket" vector in V.

Example 3.32 Examples of inner product spaces.

- The vector space $\mathbb{R}^n$, with the familiar "dot" product given by

$$\langle \mathbf{v}, \mathbf{w} \rangle := (v_1, \ldots, v_n) \cdot (w_1, \ldots, w_n) = v_1 w_1 + \cdots + v_n w_n.$$

- The vector space $\mathbb{C}^n$, with the Hermitian inner product given by

$$\langle \mathbf{v}, \mathbf{w} \rangle := v_1 \overline{w}_1 + \cdots + v_n \overline{w}_n,$$

  where the bar denotes the complex conjugate.
- Let $C[a, b]$ be the vector space of real-valued continuous functions on the interval $[a, b] \subset \mathbb{R}$. Then for $f, g \in C[a, b]$, the quantity

$$\langle f, g \rangle := \int_a^b f(x)g(x)\,dx$$

  defines an inner product on $C[a, b]$.
- Let $\mathcal{P}$ denote the vector space of polynomials in n variables with real coefficients. Then an inner product is defined by

$$\langle p, q \rangle := p(\partial)q(\mathbf{x})|_{\mathbf{x}=0}.$$

Here $p(\partial)$ means replace the variables by partial derivatives. In $\mathcal{P}(x, y)$ for example, $p(x, y) = x^2 y$ becomes $p(\partial) = \frac{\partial^2}{\partial x^2} \frac{\partial}{\partial y}$, and then apply this to the polynomial $q(x, y)$ and evaluate the result at $(x, y) = (0, 0)$. Note that the monomials are orthogonal with respect to this inner product.
- Let $\mathrm{Mat}_{m,n}(\mathbb{R})$ be the vector space of all real $m \times n$ matrices, with inner product defined by

$$\langle A, B \rangle := \mathrm{tr}(A^T B).$$

- Let $L^2([-\pi, \pi])$ be the space of integrable functions on the interval $[-\pi, \pi] \subset \mathbb{R}$ such that

$$\int_{-\pi}^{\pi} f^2 \, dx < \infty.$$

Then

$$\langle f, g \rangle := \frac{1}{\pi} \int_{-\pi}^{\pi} f(x)g(x) \, dx$$

defines an inner product on $L^2([-\pi, \pi])$.
- The *Fock space* of entire[3] functions on $\mathbb{C}^n$ that satisfy

$$\int_{\mathbb{C}^n} f(z)\overline{f(z)}e^{-z \cdot \bar{z}} \, dz < \infty,$$

with Hermitian inner product given by

$$\langle f, g \rangle := \frac{1}{\pi^n} \int_{\mathbb{C}^n} f(z)\overline{g(z)}e^{-z \cdot \bar{z}} \, dz.$$

Here we identify $\mathbb{C}^n$ with $\mathbb{R}^{2n}$. This space has important applications to quantum mechanics.
- Let G be a finite group, and denote by $\mathbb{C}[G]$ the vector space of all functions from G to $\mathbb{C}$. For α and β in $\mathbb{C}[G]$, the formula

$$\langle \alpha, \beta \rangle := \frac{1}{|G|} \sum_{g \in G} \alpha(g)\overline{\beta(g)}$$

defines an inner product on $\mathbb{C}[G]$. We will use this inner product later.

Remark 3.33 Here are some examples of normed vector spaces. Let $V = \mathbb{R}^n$, and let $\mathbf{v} = (v_1, \ldots, v_n)$.

(1) The *2-norm* is the usual Euclidean norm given by $\|\mathbf{v}\|_2 = \left(\sum_{i=1}^{n} (v_i)^2\right)^{1/2}$, which is inherited form the usual inner product.
(2) The *1-norm*; $\|\mathbf{v}\|_1 := \sum_{i=1}^{n} |v_i|$.
(3) The *p-norm* for any integer $p \geq 1$; $\|\mathbf{v}\|_p := \left(\sum_{i=1}^{n} |v_i|^p\right)^{1/p}$.
(4) The *supremum-norm*; $\|\mathbf{v}\|_\infty := \max_i |v_i|$.

---

[3] Entire functions are differentiable on the entire complex plane.

These constructions can be extended to function spaces. For example, if $f \in C[a, b]$ as in Example 3.32, then we can define

$$\|f\|_p := \left(\int_a^b |f|^p \right)^{1/p}.$$

Exercise 3.39 It is sufficient to check this for the basis monomials, and note that $g.\mathbf{v} = g.\mathbf{w}$ if and only if $\mathbf{v} = \mathbf{w}$.

Exercise 3.40 If $\mathbf{w} \in W$ and $\mathbf{u} \in W^\perp$, then

$$\langle g.\mathbf{u}, \mathbf{w} \rangle = \langle g^{-1}.[g.\mathbf{u}], g^{-1}.\mathbf{w} \rangle = \langle \mathbf{u}, g^{-1}.\mathbf{w} \rangle = 0,$$

first by G-invariance of the inner product and then by the G-invariance of W. Hence $g.\mathbf{u} \in W^\perp$, so $W^\perp$ is G-invariant.

Here are two ways show that $V = W \oplus W^\perp$. If V is finite-dimensional, let $\{\mathbf{w}_1, \ldots, \mathbf{w}_k\}$ be a basis for W, which we can extend to a basis $\{\mathbf{w}_1, \ldots, \mathbf{w}_k, \mathbf{w}_{k+1}, \ldots \mathbf{w}_n\}$ for V. Next, apply Gram-Schmidt orthonormalization to this basis for V to obtain an orthonormal basis $\{\mathbf{u}_1, \ldots \mathbf{u}_n\}$ for which $\{\mathbf{u}_1, \ldots \mathbf{u}_k\}$ will be a basis for W, and $\{\mathbf{u}_{k+1}, \ldots \mathbf{u}_n\}$ will be a basis for $W^\perp$.

Alternatively, let P_W be the orthogonal projection onto W. Then $W^\perp = \ker P$, and $V = \text{Im } P \oplus \text{Ker } P = W \oplus W^\perp$ by Exercise 2.27.

Exercise 3.43 To show that we have an inner product is a routine, if somewhat tedious, verification of the definitions involved. To show G-invariance, that $\langle h.\mathbf{v}, h.\mathbf{w} \rangle' = \langle \mathbf{v}, \mathbf{w} \rangle'$ for any $h \in G$, use the fact that summing over all $g \in G$ is the same as summing over gh for all $g \in G$ and a fixed h.

Remark 3.44 Informally, a set is complete if there are no "holes" in it. For example, the rational numbers $\mathbb{Q}$ are not complete because there are lots of "holes" such as π, $\sqrt{2}$, etc. More formally, a set is complete if every "nicely convergent" sequence (called a *Cauchy sequence*) converges to an element in that set. For example, $(3, 3.1, 3.14, 3.141, \ldots)$ is a sequence of rational numbers that converges to π which is not rational, so $\mathbb{Q}$ is not complete. The real and complex numbers are complete. Of course, in order to talk about convergence we need some notion of "distance" (called a *metric*) so that expressions such as "$|a_n - L| < \epsilon$" make sense. For a Hilbert space, we use the norm obtained from the inner product.

Normed vector spaces (without an inner product) that are complete are called *Banach spaces*, (after the Polish mathematician Stefan Banach), and play a central role in functional analysis.

The space $C[a, b]$ in the examples is not a Hilbert space since there are sequences of continuous functions that converge (in the norm defined by the inner product) to a function that is not continuous (Exercise 3.45), so the space is not complete.

One of the more straightforward examples of a Hilbert space is the vector space $\ell^2(\mathbb{R})$ of square-summable sequences in $\mathbb{R}$. In other words, vectors are sequences of the form $(a_1, a_2, \ldots)$ such that $\sum_{i=1}^{\infty} a_i^2$ exists (is finite), and with vector addition and scalar multiplication defined component-wise, as in $\mathbb{R}^n$. Here the inner product is defined as a sort of "infinite dot product"

$$\langle (a_1, a_2, \ldots), (b_1, b_2, \ldots) \rangle := \sum_{i=1}^{\infty} a_i b_i.$$

The Cauchy-Schwartz inequality guarantees that this sum is finite, so the inner product is well defined. The set of standard basis vectors $\{e_i\}_{i=1}^{\infty}$ is an orthonormal basis for this space, in the sense that we can write any vector in $\ell^2(\mathbb{R})$ as an infinite linear combination of basis vectors that converges in the norm obtained from the inner product. Sometimes the term *orthonormal system* is used instead of *basis* in this context. Proving that this space is complete is a good exercise for the motivated reader with an interest in analysis. What do we mean by a "convergent sequence of elements in $\ell^2(\mathbb{R})$?"

The space $L^2([-\pi, \pi])$ from Example 3.32 is also a Hilbert space with basis

$$\{1/2, \sin(nx), \cos(nx) \mid n = 1, 2, 3 \ldots\}.$$

Bases for less well-behaved infinite-dimensional vector spaces can be much more complicated. See [Sm] for a nice discussion that should be accessible to most of this audience.

Exercise 3.45 It is a routine calculus exercise that to show that the sequence f_n *converges point-wise* to f. That is, for each x in the interval $[0, 1]$, the sequence of numbers $\{f_n(x)\}$ converges to $f(x)$.

To show convergence with respect to the norm obtained from the inner product, we have

$$\| f_n - f \| = \langle f_n - f, f_n - f \rangle = \langle f_n, f_n \rangle - 2 \langle f_n, f \rangle + \langle f, f \rangle.$$

Considering each of the terms:

$$\langle f_n, f_n \rangle = \int_0^1 x^n \cdot x^n \, dx + \int_1^1 1 \cdot 1 \, dx = \frac{1}{2n+1},$$

$$\langle f_n, f \rangle = \int_0^1 x^n \cdot 0 \, dx + \int_1^1 x^n \cdot 1 \, dx = 0,$$

$$\langle f, f \rangle = \int_0^1 0 \cdot 0 \, dx + \int_1^1 x^n \cdot 1 \, dx = 0.$$

Thus $\|f_n - f\| = \frac{1}{2n+1}$ and so $f_n \to f$ in the norm on $C[0, 1]$.

For the record, there are other norms on $C[0, 1]$ for which this sequence does not converge, and for which this space is complete, although no longer an inner product space.

Exercise 3.47 The observation that the map $\phi_{\mathbf{w}}(\mathbf{v}) := \langle \mathbf{v}, \mathbf{w} \rangle$ defines a linear functional on any inner-product space means that we can exploit Example 3.32 for examples of linear functionals.

If $f, g \in C([a, b])$ then

$$\phi_1(f) = \int_a^b f(x)dx, \quad \text{and} \quad \phi_g(f) = \int_a^b f(x)g(x)dx$$

are both linear functionals, as is the *evaluation functional*;

$$e_c(f) := f(c) \text{ for } c \in [a, b].$$

The Structure of the Symmetric Group

4

In this chapter we delve further into properties of the symmetric group, some of which you were asked to prove in previous chapters. We also introduce the notion of conjugation and conjugacy classes, and classify elements of the symmetric group according to conjugacy classes that correspond to their presentation as a product of disjoint cycles.

4.1   Cycles and Cycle Structure

Proposition 4.1 *If $k \in \{1, 2, \ldots, n\}$ and $\sigma \in S_n$, then there is a positive integer m such that $\sigma^{m+1}.k = k$, and the elements $\{k, \sigma.k, \sigma^2.k, \ldots, \sigma^m.k\}$ are all distinct.*

Proof Consider the set $\{k, \sigma.k, \sigma^2.k, \ldots\}$. Since σ is a bijection of the finite set $\{1, 2, \ldots, n\}$, there must be some positive integers r and t (we may assume $r > t$) so that $\sigma^r.k = \sigma^t.k$. Thus $\sigma^r (\sigma^t)^{-1}.k = \sigma^{r-t}.k = k$. There are lots of possibilities for r and t, but those that yield the smallest value for $r-t$ guarantee that the elements in $\{k, \sigma.k, \sigma^2.k, \ldots, \sigma^m.k\}$ are all distinct, where $m = r - t - 1$.   □

Proposition 4.2 *Every permutation $\sigma \in S_n$ can be written as a product of disjoint cycles.*

For example, $(1, 2, 3, 4)(1, 5, 3) = (1, 5, 4)(2, 3)$ and $(1, 3, 4)(3, 4, 5) = (1, 3)(4, 5)$. You should work out more examples.

Proof Choose some $k \in \{1, 2, \ldots, n\}$, let m be as in Proposition 4.1, and consider the cycle

$$(k, \sigma.k, \sigma^2.k, \ldots, \sigma^m.k).$$

If $m = n$, then σ consists of just one cycle of length n and we are done. If not, then there is some $b \in \{1, 2, \ldots, n\}$ that is not in the above cycle, and we can repeat the above procedure with b. The cycles must be disjoint because if $\sigma^r.k = \sigma^t.b$ for some $1 \le t < r \le n$, then $\sigma^{r-t}.k = b$, contradicting the choice of b. Continue by induction until all the elements of $\{1, 2, \ldots, n\}$ are in a cycle.   □

Proposition 4.3 *Disjoint cycles commute.*

Proof Let σ and τ represent two disjoint cycles in S_n, and choose some arbitrary $j \in \{1, 2, \ldots, n\}$. Since σ and τ are disjoint, at least one of them, say τ, fixes j. But then τ must also fix $\sigma.j$ because the cycles are disjoint. Hence $\sigma\tau.j = \sigma.j = \tau\sigma.j$. If both σ and τ fix j, then $\sigma\tau.j = \sigma.j = j$ and $\tau\sigma.j = \tau.j = j$.   □

We can classify the permutations in S_n according to the lengths of their disjoint cycles. For example, $(1, 2, 3, 4)(5, 6)(7, 8)(9, 10, 11) \in S_{11}$ has one 4-cycle, two 2-cycles and one 3-cycle, and so has *cycle structure* $(4, 2, 2, 3)$. Of course $(1, 2)(3, 4, 5)(6, 7, 8, 9)(10, 11)$ has the same cycle structure, so it makes sense to describe a cycle structure in some unique way, such as in non-increasing order: $(4, 3, 2, 2)$ in this example. That is, we can label cycle structures in S_n by partitions of n. Furthermore, using the cycle class notation from Sect. 2.3, the above example has cycle class $(1^0, 2^2, 3^1, 4^1)$ and signature $(2, 1, 1, 0)$.

Exercise* 4.4 How many elements in S_n have a given cycle class? Hint: Mimic the reasoning for Exercise 2.13.

Exercise 4.5 Let $\tau = (t_1, t_2, \ldots, t_{k-1}, t_k)$ be a cycle in S_n. Show that

$$\tau^{-1} = (t_1, t_k, t_{k-1}, \ldots, t_2).$$

Hint: Where does τ send some arbitrary t_j? Conclude that for any $\sigma \in S_n$, σ and σ^{-1} have the same cycle structure.

4.2   Generators and Parity

The following proposition can often simplify proofs involving the symmetric group.

Proposition 4.6 *The group S_n is generated by the two-cycles, also called* transpositions. *That is, every permutation can be written as a finite product of (not necessarily disjoint) transpositions.*

Proof Since any permutation can be written as a product of disjoint cycles, it is sufficient to prove this for an arbitrary cycle. Show by induction on r that $(a_1, a_2, a_3, \ldots, a_r) = (a_1, a_2)(a_2, a_3) \cdots (a_{r-1}, a_r)$.   □

Exercise 4.7 Write out some examples that illustrate the above proposition.

Definition 4.8 If $\sigma \in S_n$ can be written as the product of an even number of two-cycles, then we say it is an *even permutation*, otherwise it is an *odd permutation*. A permutation can be written as a product of two-cycles in multiple ways, but the *parity* (whether the number of these transpositions is even or odd) **is** unique.

Example 4.9 $(2, 3) = (1, 2)(1, 3)(1, 2) = (1, 3)(1, 4)(2, 4)(1, 4)(1, 3)$. Again, you should write out more examples on your own.

Proposition 4.10 (The Parity Theorem) *A permutation cannot be both even and odd.*

There are a number of proofs of this; [G], Theorem 5.5, and [L1], Chapter 1, Exercise 30 (a) are two completely different proofs. The proof we present here can be found in [JK], Lemma 1.1.19.

Proof The result follows from that fact that the function

$$\mathrm{sgn}(\sigma) := \begin{cases} 1, & \text{if } \sigma \text{ is even;} \\ -1, & \text{if } \sigma \text{ is odd,} \end{cases}$$

is a well-defined group homomorphism from S_n to the multiplicative group $\{-1, 1\}$. This function is called the *sign* or *signum* function, so as not to be confused with the trigonometric sine function. For $n \geq 2$, we define the *difference product* $\triangle_n$ by

$$\triangle_n := \prod_{1 \leq i < j \leq n} (j - i) \in \mathbb{Z},$$

and define an action of $\sigma \in S_n$ on $\triangle_n$ by

$$\sigma \triangle_n := \prod_{1 \leq i < j \leq n} [\sigma(j) - \sigma(i)].$$

Now $\sigma \triangle_n$ is obviously well-defined since each $\sigma(j)$ and $\sigma(i)$ are well-defined. Since σ is a bijection of the set $\{1, 2, \ldots, n\}$, the action of σ leaves the factors of $\triangle_n$ unchanged except as to sign, and hence $\sigma \triangle_n = \pm \triangle_n$. We can thus define $\mathrm{sgn}(\sigma)$, the *sign* of σ, by

$$\mathrm{sgn}(\sigma) := \frac{\sigma \triangle_n}{\triangle_n} = \frac{\prod_{1 \leq i < j \leq n}(\sigma(j) - \sigma(i))}{\prod_{1 \leq i < j \leq n}(j - i)} = \prod_{1 \leq i < j \leq n} \frac{\sigma(j) - \sigma(i)}{j - i} \in \{-1, 1\}.$$

It remains to show that $\text{sgn} \colon S_n \to (\{-1, 1\}, \times)$ is a group homomorphism, *i.e.*, that $\text{sgn}(\sigma\tau) = \text{sgn}(\sigma)\,\text{sgn}(\tau)$. To wit:

$$
\begin{aligned}
\text{sgn}(\sigma\tau) &= \prod_{1 \le i < j \le n} \frac{\sigma\tau(j) - \sigma\tau(i)}{j - i} \\[2mm]
&= \prod_{1 \le i < j \le n} \frac{\sigma\tau(j) - \sigma\tau(i)}{j - i} \, \frac{\tau(j) - \tau(i)}{\tau(j) - \tau(i)} \\[2mm]
&= \prod_{1 \le i < j \le n} \frac{\sigma\tau(j) - \sigma\tau(i)}{\tau(j) - \tau(i)} \, \frac{\tau(j) - \tau(i)}{j - i} \\[2mm]
&= \prod_{1 \le i < j \le n} \frac{\sigma\tau(j) - \sigma\tau(i)}{\tau(j) - \tau(i)} \prod_{1 \le i < j \le n} \frac{\tau(j) - \tau(i)}{j - i} \\[2mm]
&\doteq \text{sgn}(\sigma)\,\text{sgn}(\tau).
\end{aligned}
$$

Now note that if σ is a transposition, then $\sigma \triangle_n = -\triangle_n$, and hence $\text{sgn}(\sigma) = -1$. Finally, since sgn is a group homomorphism, it follows that $\text{sgn}(\sigma) = 1$ if σ can be written as a product of an even number of transpositions, and that $\text{sgn}(\sigma) = -1$ if σ can be written as a product of an odd number of transpositions.   □

Exercise 4.11 Write out $\triangle_n$ and $\sigma.\triangle_n$ for several values of n and for several permutations σ.

Exercise* 4.12 Show that if σ is a transposition, then $\sigma \triangle_n = -\triangle_n$.

Exercise 4.13 Determine the parity of the elements in S_2, S_3 and S_4. You do not need to write out all the elements in each group, just one example from each cycle class.

Exercise 4.14 Show that the set of even permutations in S_n is a subgroup of S_n, called the *alternating group*, denoted A_n. Note that A_n is the kernel of the sgn homomorphism.

Exercise 4.15 This exercise outlines another proof of the Parity Theorem that uses facts from linear algebra.

(1) For $\sigma \in S_n$, let $\widetilde{\sigma}$ be its associated permutation matrix. See Exercise 1.86.
(2) Recall that the map $\sigma \mapsto \widetilde{\sigma}$ a group homomorphism from S_n to $GL(n, \mathbb{R})$.
(3) Show that if (i, j) is a transposition, then the matrix realization $\widetilde{(i, j)}$ is obtained by interchanging the ith and jth columns of the identity matrix.
(4) Recall that the map from $GL(n, \mathbb{R})$ to $\mathbb{R}$ given by $A \mapsto \text{Det}(A)$ is a group homomorphism, and hence the composition $\sigma \mapsto \widetilde{\sigma} \mapsto \text{Det}(\widetilde{\sigma})$ is a group homomorphism from S_n to $\mathbb{R}$.

(5) Recall the "minor expansion formula" for the determinant of a matrix, and what happens to the determinant when we interchange two columns of the matrix. Conclude that $\text{Det}(\widetilde{\sigma}) = +1$ if σ is an even permutation, and $\text{Det}(\widetilde{\sigma}) = -1$ if σ is odd.

4.3 Conjugation and Conjugacy Classes

Recall from linear algebra that two square matrices A and B are *similar* if there is an invertible matrix P such that $B = PAP^{-1}$. We have an analogous construction for elements in a group.

Definition 4.16 Let G be a group, and let $h \in G$. The map $\phi_h : G \to G$ defined by $\phi_h(g) = hgh^{-1}$ is called *conjugation by* h. Given two elements g and g' in G, we say that g *is conjugate to* g' if there is some $h \in G$ such that $g' = hgh^{-1}$.

Exercise* 4.17 Show that conjugation by h is an automorphism of G.

Exercise 4.18 Show that conjugacy is an equivalence relation. Consequently, any group G can be written as a disjoint union of distinct conjugacy classes. Hint: Mimic the proof from linear algebra, that similarity of matrices is an equivalence relation. Also show that the conjugacy classes are the orbits of the group action given in Example 1.21.

Exercise 4.19 Show that if $\tau = (t_1, t_2, \ldots, t_k)$ is a cycle in S_n, then for any $\sigma \in S_n$ $\sigma\tau\sigma^{-1} = (\sigma.t_1, \sigma.t_2, \ldots, \sigma.t_k)$. Hint: Check that $(\sigma\tau\sigma^{-1}).\sigma.t_i = \sigma.t_{i+1}$.

Applying Exercise 4.19 and Proposition 4.2 yields a particularly nice way of labeling conjugacy classes for the symmetric group.

Proposition 4.20 *Two elements in S_n are conjugate if and only if they have the same cycle structure. As a consequence, we can label the conjugacy classes in S_n by the partitions of n.*

Exercise 4.21 How many elements are in each conjugacy class? Hint: Exercise 4.4

We can also conjugate subgroups of a group.

Definition 4.22 Let H and K be two subgroups of a group G. Then H *is conjugate to K in G* if, for some $g \in G$, $H = gKg^{-1} := \{gkg^{-1} \mid k \in K\}$. Note that conjugacy is also an equivalence relation on the set of subgroups of G.

4.4   Hints and Additional Comments

Exercise 4.4 It is usually a good idea to first try some examples. Consider the permutation $(1, 2)(3, 4)(5, 6)(7, 8, 9)(10, 11, 12) \in S_{12}$. While there are 12! ways to reorder the numbers $1, \ldots, 12$, some will result in the same permutation. There are 3! ways to permute the three two-cycles and 2! ways to permute the two three-cycles. Also, we can cyclically permute each of the three-cycles in three ways, *i.e*, $(7, 8, 9) = (8, 9, 7) = (9, 7, 8)$, and each of the two-cycles in two ways. Thus there are $12!/(3! 2^3\, 2!\, 3^2)$ distinct permutations with this same cycle-class.

Generalizing this, we see that if a permutation in S_n has cycle class $(1^{\lambda_1}, 2^{\lambda_2} \ldots, k^{\lambda_k})$, then the number of distinct permutations with the same cycle-class is

$$\frac{n!}{1^{\lambda_1}\lambda_1!\, 2^{\lambda_2}\lambda_2! \ldots k^{\lambda_k}\lambda_k!}.$$

The denominator in this expression appears often enough that it is given a special designation by some authors ([M] pg 24, [Sa] pg. 3). If $\lambda \vdash n$ and if $m_i = m_i(\lambda)$ is the number of parts of λ equal to i (Sect. 2.3), then

$$z_\lambda := \prod_{i \geq 1} i^{m_i} m_i! \,.$$

If $g \in S_n$ has cycle type λ, then the number z_λ is the order of the subgroup $Z_g := \{h \in S_n \mid hgh^{-1} = g\}$, called the *centralizer* of $g \in S_n$. While you're at it, you should verify that if G is any group, and if $g \in G$, then Z_g is in fact a subgroup of G.

Exercise 4.12 Observe that if $\sigma = (i, j)$ with $i < j$, then σ only affects the following factors in Δ_n:

(1) $(j - i)$ for $i < j$,
(2) $(i - k), (j - k)$ for $k < i < j$,
(3) $(\ell - i), (\ell - j)$ for $i < j < \ell$,
(4) $(m - i), (j - m)$ for $i < m < j$.

The effect of σ is that the factor in (1) changes sign, the factors in (2) and (3) are unchanged except as to the order of the two factors, and both factors in (4) change sign. The result follows.

Exercise 4.17 First note that, for all $h, g_1, g_2 \in G$,

$$\phi_h(g_1 g_2) = h\,(g_1\, g_2)\, h^{-1} = (h\, g_1\, h^{-1})\,(h\, g_2\, h^{-1}) = \phi_h(g_1)\phi_h(g_2),$$

so $\phi \colon G \to G$ is a homomorphism. Now verify that ϕ is a bijection. What is $(\phi_h)^{-1}$?

S_n-Decomposition of Polynomial Spaces for $n = 1, 2, 3$

5

Towards the goal of providing examples, in this chapter we demonstrate explicit decompositions of some homogeneous polynomial spaces $\mathcal{P}_k(x_1, \ldots, x_n)$ into S_n-irreducible subspaces for $n = 1, 2, 3$. At this stage the methods we use are mostly ad hoc, and generalizing these results for $n > 3$ will require more sophisticated tools than those which we employ here.

Notation 5.1 From now on it will be convenient to denote a vector space V with basis $\{\mathbf{v}_1, \mathbf{v}_2, \ldots, \mathbf{v}_n\}$ as $V = \langle\!\langle \mathbf{v}_1, \mathbf{v}_2, \ldots, \mathbf{v}_n \rangle\!\rangle$, or $V = \langle\!\langle \mathbf{v}_1, \mathbf{v}_2, \ldots, \mathbf{v}_n \rangle\!\rangle_{\mathbb{F}}$ if it is necessary to specify the field of scalars $\mathbb{F}$.

5.1 S_1

For the sake of completeness, and at the risk of stating the obvious, the group S_1 consists of only the identity element. The definition of a group action implies that the only representation of S_1 is the *trivial* or *identity* representation, given by $\sigma.\mathbf{v} = \mathbf{v}$ for all $\mathbf{v} \in V$ and all $\sigma \in S_1$. Any k-dimensional vector space decomposes into a direct sum of k copies of one-dimensional trivial representations of S_1.

5.2 S_2

Since $S_2 = \{(1), (1, 2)\}$, any action of S_2 on a polynomial $\mathbf{p} \in \mathcal{P}(x, y)$ will either leave $\mathbf{p}$ unchanged or switch the x and y.

Similar to Example 2.29, $\mathcal{P}_1(x, y) = \langle\!\langle \mathbf{x}, \mathbf{y} \rangle\!\rangle$ certainly contains a one-dimensional trivial representation $\mathcal{I}_2 = \langle\!\langle\, \mathbf{x} + \mathbf{y}\, \rangle\!\rangle$. The complementary subspace, which will be more appropriate to call $\mathcal{A}_2$ in this case, is therefore one dimensional with an obvious basis $\mathcal{A} = \langle\!\langle \mathbf{x} - \mathbf{y} \rangle\!\rangle$. Note that, for any $\mathbf{v} \in \mathcal{A}_2$ and any $\sigma \in S_2$, we have $\sigma.\mathbf{v} = \pm\mathbf{v}$, depending on whether $\sigma = (1)$ (an even permutation), or $\sigma = (1, 2)$

© The Author(s), under exclusive license to Springer Nature Switzerland AG 2022
R. M. Howe, *An Invitation to Representation Theory*, SUMS Readings,
https://doi.org/10.1007/978-3-030-98025-2_5

(an odd permutation). This is the simplest example of the *alternating representation* or the *sign representation*, given by $\sigma.\mathbf{v} = \text{sgn}(\sigma)\mathbf{v}$. The alternating representation is one-dimensional and so irreducible. To summarize, $\mathcal{P}_1(x, y) \cong \mathcal{I} \oplus \mathcal{A}$.

Notation 5.2 If we need to distinguish between the trivial, alternating, etc. representations of S_n for *different n*, we will write $\mathcal{I}_n$, $\mathcal{A}_n$, etc.

Next we have $\mathcal{P}_2(x, y) = \mathcal{P}_{(2,0)} \oplus \mathcal{P}_{(1,1)}$. Since $\mathcal{P}_{(2,0)} = \langle\!\langle \mathbf{x}^2, \mathbf{y}^2 \rangle\!\rangle$, it should be apparent that $\mathcal{P}_{(2,0)} \cong \mathcal{I}_2 \oplus \mathcal{A}_2 \cong \mathcal{P}_{(1,0)}$. Since $\mathcal{P}_{(1,1)} = \langle\!\langle \mathbf{xy} \rangle\!\rangle$, it should also be apparent that $\mathcal{P}_{(1,1)}$ consists of only the identity representation. Thus $\mathcal{P}_2(x, y) \cong \mathcal{I}_2 \oplus \mathcal{I}_2 \oplus \mathcal{A}_2$.

So far we have only seen two irreducible polynomial representations of S_2, the trivial and alternating representations denoted $\mathcal{I}_2$ and $\mathcal{A}_2$; we now show that this is all of them. Let $f(x, y)$ be any function of two variables. Define $f_s(x, y) := f(x, y) + f(y, x)$, which is said to be *symmetric* since it is unchanged by switching x and y. Define $f_a(x, y) := f(x, y) - f(y, x)$, which is said to be *alternating* since it changes sign when switching x and y. Then $f(x, y) = \frac{1}{2}[f_s(x, y) + f_a(x, y)]$. That is, any function in x and y (and hence any polynomial) can be written as a sum of a symmetric function and an alternating function. Consequently, any S_2-invariant subspace of $\mathcal{P}(x, y)$ decomposes into a direct sum of a number of copies of the one-dimensional trivial and alternating representations of S_2.

Exercise 5.3 Decompose $\mathcal{P}_k(x, y)$ into S_2-irreducible subspaces for several values of $k \geq 3$.

Exercise 5.4 Show that the maps $f \mapsto \frac{1}{2}f_s$ and $f \mapsto \frac{1}{2}f_a$ are projections. Describe the subspaces onto which they project.

5.3 S_3

We have seen in Sect. 2.6 that $\mathcal{P}_1 = \mathcal{P}_{(1,0,0)} = \langle\!\langle \mathbf{x}, \mathbf{y}, \mathbf{z} \rangle\!\rangle$ decomposes as the direct sum of two irreducible subspaces:

$$\mathcal{P}_1 \cong \mathcal{I}_3 \oplus \mathcal{W}_3.$$

The identity representation has an obvious basis, $\mathcal{I}_3 = \langle\!\langle \mathbf{x} + \mathbf{y} + \mathbf{z} \rangle\!\rangle$, and it is clearly irreducible since it's one dimensional. The complementary subspace

$$\mathcal{W}_3 = \{r\mathbf{x} + s\mathbf{y} + t\mathbf{z} \mid r, s, t \in \mathbb{C}; r + s + t = 0\}$$

is irreducible (Exercise 2.31) and two dimensional, so there are lots of choices for a basis. One choice would be

$$\mathcal{W}_3 = \langle\!\langle \mathbf{x} - \mathbf{y}, \mathbf{x} - \mathbf{z} \rangle\!\rangle.$$

Now $\mathcal{P}_2 = \mathcal{P}_{(2,0,0)} \oplus \mathcal{P}_{(1,1,0)}$, and it's easy to see that $\mathcal{P}_{(2,0,0)} = \langle\!\langle \mathbf{x}^2, \mathbf{y}^2, \mathbf{z}^2 \rangle\!\rangle$ decomposes just like $\mathcal{P}_{(1,0,0)}$;

$$\mathcal{P}_{(2,0,0)} \cong \mathcal{I}_3 \oplus \mathcal{W}_3.$$

In this case $\mathcal{I}_3 = \langle\!\langle \mathbf{x}^2 + \mathbf{y}^2 + \mathbf{z}^2 \rangle\!\rangle$, and $\mathcal{W}_3 = \langle\!\langle \mathbf{x}^2 - \mathbf{y}^2, \mathbf{x}^2 - \mathbf{z}^2 \rangle\!\rangle$. The map defined on the basis elements by

$$\begin{cases} \mathbf{x} \mapsto \mathbf{x}^2, \\ \mathbf{y} \mapsto \mathbf{y}^2, \\ \mathbf{z} \mapsto \mathbf{z}^2. \end{cases}$$

is an S_3-map from $\mathcal{P}_{(1,0,0)}$ to $\mathcal{P}_{(2,0,0)}$ that is an S_3-isomorphism between the irreducible subspaces. It's not quite so obvious that $\mathcal{P}_{(1,1,0)}$ also decomposes "just like" $\mathcal{P}_1$, although we might get a hint since $\mathcal{P}_{(1,1,0)} = \langle\!\langle \mathbf{xy}, \mathbf{xz}, \mathbf{yz} \rangle\!\rangle$ is three dimensional. We again have a copy of

$$\mathcal{I}_3 = \langle\!\langle \mathbf{xy} + \mathbf{xz} + \mathbf{yz} \rangle\!\rangle,$$

with complimentary subspace

$$\mathcal{W}_3 = \{ r\,\mathbf{xy} + s\,\mathbf{xz} + t\,\mathbf{yz} \mid r, s, t \in \mathbb{C}, r + s + t = 0 \}.$$

Exercise* 5.5 Find G-maps from $\mathcal{P}_1$ to $\mathcal{P}_{(1,1,0)}$. There are several possibilities.

We know that $\mathcal{P}_3 = \mathcal{P}_{(3,0,0)} \oplus \mathcal{P}_{(2,1,0)} \oplus \mathcal{P}_{(1,1,1)}$. It should be clear by now that $\mathcal{P}_{(3,0,0)} = \langle\!\langle \mathbf{x}^3, \mathbf{y}^3, \mathbf{z}^3 \rangle\!\rangle \cong \mathcal{I}_3 \oplus \mathcal{W}_3$ and that $\mathcal{P}_{(1,1,1)} = \langle\!\langle \mathbf{xyz} \rangle\!\rangle \cong \mathcal{I}_3$.

We might begin to suspect that all the polynomial representations of S_3 decompose into copies of $\mathcal{I}_3$ and $\mathcal{W}_3$. We might also begin to suspect that these are the only irreducible representations of S_3. The problem here is that the polynomial spaces we've considered up to now haven't been "big" enough, that is, of sufficiently large dimension. Things get more interesting when we decompose the 6-dimensional space $\mathcal{P}_{(2,1,0)} = \langle\!\langle \mathbf{x}^2\mathbf{y}, \mathbf{x}^2\mathbf{z}, \dots, \mathbf{yz}^2 \rangle\!\rangle$. We certainly have a copy of $\mathcal{I}_3 = \langle\!\langle \mathbf{x}^2\mathbf{y} + \mathbf{x}^2\mathbf{z} + \dots + \mathbf{yz}^2 \rangle\!\rangle$, and we have also seen (Exercise 3.20) that $\mathcal{P}_{(2,1,0)}$ contains two copies of $\mathcal{W}_3$. By adding up the dimensions, it follows that there must be another one-dimensional subrepresentation. The missing representation turns out to be the *alternating representation* $\mathcal{A}_3$, with the group action given by $\sigma.\mathbf{v} = \mathrm{sgn}(\sigma)\mathbf{v}$.

We can find a basis for $\mathcal{A}_3$ using brute force as follows: consider a linear combination of the monomials in $\mathcal{P}_{(2,1,0)}$ of the form $\mathbf{v} = r_1\mathbf{x}^2\mathbf{y} + r_2\mathbf{x}^2\mathbf{z} + \dots + r_6\mathbf{yz}^2$. Since $\mathbf{v}$ changes sign when acted on by the transposition $(1, 2)$ for example, the monomials $\mathbf{x}^2\mathbf{y}$ and $\mathbf{xy}^2$ must have opposite coefficients in this sum, and similarly for the other transpositions acting on the monomials. Thus $\mathcal{A}_3$ is spanned by the vector

$$\mathbf{v} = \mathbf{x}^2\mathbf{y} - \mathbf{x}\mathbf{y}^2 - \mathbf{x}^2\mathbf{z} + \mathbf{x}\mathbf{z}^2 + \mathbf{y}^2\mathbf{z} - \mathbf{y}\mathbf{z}^2.$$

Exercise* 5.6 Verify that $\sigma.\mathbf{v} = \mathrm{sgn}(\sigma)\mathbf{v}$ for all $\sigma \in S_3$ and $\mathbf{v} \in \mathcal{A}_3$ as above. Hint: You only need to check this for the two-cycles. Why?

Exercise 5.7 Check that $\mathbf{v} = \sum_{\tau \in S_3} \mathrm{sgn}(\tau)\tau.\mathbf{x}^2\mathbf{y}$. Use this for another proof that $\sigma.\mathbf{v} = \mathrm{sgn}(\sigma)\mathbf{v}$.

Summarizing, we have that $\mathcal{P}_{(2,1,0)} \cong \mathcal{I}_3 \oplus \mathcal{W}_3 \oplus \mathcal{W}_3 \oplus \mathcal{A}_3$. We say the $\mathcal{W}_3$ appears in the decomposition of $\mathcal{P}_{(2,1,0)}$ with *multiplicity two*, which is written $\mathcal{P}_{(2,1,0)} \cong \mathcal{I}_3 \oplus 2\mathcal{W}_3 \oplus \mathcal{A}_3$, or $\mathcal{P}_{(2,1,0)} \cong \mathcal{I}_3 \oplus \mathcal{W}_3^{\oplus 2} \oplus \mathcal{A}_3$. More generally, if U is a vector space and $k \in \mathbb{N}$, the notation kU or $U^{\oplus k}$ means $\underbrace{U \oplus \cdots \oplus U}_{k\ summands}$. If a representation V decomposes into subrepresentations

$$V \cong kU \oplus W,$$

and if W contains no subrepresentations equivalent to U, then we say k is the *multiplicity of U in V*, or that U *appears in V with multiplicity k*.

Exercise* 5.8 Decompose several examples of $\mathcal{P}_k(x, y, z)$ for $k \geq 4$. Which of the $\{\mathcal{P}_\mu \mid \mu \vdash k\}$ decompose like those in $\mathcal{P}_2$ or $\mathcal{P}_3$ and why?

Remark 5.9 Polynomials such as

$$\mathbf{x} + \mathbf{y} + \mathbf{z} = \frac{1}{2} \sum_{\sigma \in S_3} \sigma.\mathbf{x} \in \mathcal{P}_{(1,0,0)},$$

and

$$\mathbf{x}^2\mathbf{y} + \mathbf{x}^2\mathbf{z} + \mathbf{x}\mathbf{y}^2 + \mathbf{x}\mathbf{z}^2 + \mathbf{y}^2\mathbf{z} + \mathbf{y}\mathbf{z}^2 = \sum_{\sigma \in S_3} \sigma.\mathbf{x}^2\mathbf{y} \in \mathcal{P}_{(2,1,0)}$$

are examples of *orbit sums*. If G is a group, any polynomial $p \in \mathcal{P}$ such that $g.p = p$ for all $g \in G$ is called an *invariant polynomial*, or a *G-invariant polynomial* if more specificity is required. Polynomials that are S_n-invariant are called *symmetric polynomials*. The polynomial $p(x, y) = x^2 + y^2$ is clearly S_2-invariant, but it is also $O(2, \mathbb{R})$-invariant (Exercise 1.57). We can define a left action of a matrix $[g] \in O(2, \mathbb{R})$ on p, (writing elements in $\mathbb{R}^2$ as column vectors) by

$$[g].p\left(\begin{bmatrix} x \\ y \end{bmatrix}\right) = p\left([g^{-1}] \begin{bmatrix} x \\ y \end{bmatrix}\right).$$

The study of symmetric and invariant polynomials, and invariant functions in general, are important areas of mathematics in their own right, and much of the representation theory of the symmetric group is closely related to the theory of symmetric functions.

Exercise 5.10 Check that orbit sums for a finite group G are G-invariant. Also verify that $p(x, y) = x^2 + y^2$ is $O(2, \mathbb{R})$-invariant.

Exercise* 5.11 Compute the Vandermonde determinant (you may need to look this up) in the variables x, y and z. Does this look familiar? What happens when two variables are interchanged?

Exercise* 5.12 A function $f(\mathbf{x}) = f(x_1, \ldots, x_n)$ is said to be *alternating* if $\sigma.f(\mathbf{x}) = \operatorname{sgn}(\sigma) f(\mathbf{x})$ for any $\sigma \in S_n$. Define

$$\triangle(\mathbf{x}) := \prod_{1 \le i < j \le n} (x_j - x_i),$$

and show that $\triangle(\mathbf{x})$ is a homogeneous polynomial of degree $n(n-1)/2$. Show that $\triangle(\mathbf{x})$ is alternating, and that any alternating polynomial $P(\mathbf{x})$ is divisible by $\triangle(\mathbf{x})$. Conclude that the Vandermonde determinant of Exercise 5.11 (also called the Vandermonde polynomial) is equal to $\triangle(\mathbf{x})$, and consequently, any alternating polynomial is divisible by the Vandermonde polynomial.

5.4   Isotypic Subspaces and Multiplicities

The direct sum of the two copies of W_3 in the decomposition $\mathcal{P}_{(2,1,0)} \cong \mathcal{I}_3 \oplus 2\,W_3 \oplus \oplus \mathcal{A}_3$ is called the W_3-*isotypic subspace* or the W_3-*isotypic component* of $\mathcal{P}_{(2,1,0)}$, that is, the direct sum of all the irreducible representations in $\mathcal{P}_{(2,1,0)}$ equivalent to W_3.

Definition 5.13 If V is a representation of a group G, and if W is an irreducible representation of G, then the W-*isotypic component* of V is the direct sum of all irreducible subrepresentations of V that are equivalent to W.

There can be numerous ways to decompose an isotypic component into it's irreducible subspaces. For example, the G-maps from Exercises 3.5 and 3.20 give explicit bases for the two distinct copies of W_3 in $\mathcal{P}_{(2,1,0)}$, but there certainly are other possibilities.

To illustrate, consider the $\mathcal{I}_3$-isotypic subspace of $\mathcal{P}_2(\mathbf{x}, \mathbf{y}, \mathbf{z}) \cong 2\mathcal{I}_3 \oplus 2W_3$. One way to decompose this would be

$$2\mathcal{I}_3 = \langle\!\langle \mathbf{x}^2 + \mathbf{y}^2 + \mathbf{z}^2 \rangle\!\rangle \oplus \langle\!\langle \mathbf{xy} + \mathbf{xz} + \mathbf{yz} \rangle\!\rangle,$$

which is inherited from the decomposition $\mathcal{P}_2 = \mathcal{P}_{(2,0,0)} \oplus \mathcal{P}_{(1,1,0)}$. But the decomposition

$$2\mathcal{I}_3 = \langle\!\langle \mathbf{x}^2 + \mathbf{y}^2 + \mathbf{z}^2 + \mathbf{xy} + \mathbf{xz} + \mathbf{yz} \rangle\!\rangle \oplus \langle\!\langle \mathbf{x}^2 + \mathbf{y}^2 + \mathbf{z}^2 - \mathbf{xy} - \mathbf{xz} - \mathbf{yz} \rangle\!\rangle$$

is equally legitimate. This is similar to the issue of choosing a basis for a 2-dimensional vector space.

More generally, if a representation V of G decomposes into a direct sum of inequivalent irreducible representations as

$$V = m_1 V_1 \oplus m_2 V_2 \oplus \cdots \oplus m_k V_k,$$

then m_i is the *multiplicity* of the irreducible representation V_i, and the subspace $m_i V_i$ is the V_i-*isotypic subspace*. Finding "natural" or "useful" ways to decompose and label multiple copies of irreducible representations can be an important problem in representation theory and its applications.

Exercise 5.14 Find different ways to decompose the isotypic subspaces in the previous examples. Find new examples.

The following exercises are generalizations of Corollaries 3.17 and 3.18, and provide a method to determine multiplicities in isotypic subspaces. Note that these results coincide with the results of Exercises 3.7–3.10.

Exercise* 5.15 Let V, U, and W be representations of a group G. Show that

$$\mathrm{Hom}_G(V, U \oplus W) \cong \mathrm{Hom}_G(V, U) \oplus \mathrm{Hom}_G(V, W).$$

Proposition 5.16 *Let V and W be two representations of a finite group G with V irreducible. Then $\mathrm{Dim}\,\mathrm{Hom}_G(V, W)$ is the multiplicity of V in W.*

Proof By complete reducibility, we can write $W \cong V_1 \oplus \cdots \oplus V_k \oplus W_1$, where each $V_i \cong V$, and where W_1 contains no subrepresentations equivalent to V. By induction on Exercise 5.15,

$$\mathrm{Hom}_G(V, W) = \mathrm{Hom}_G(V, V_1 \oplus \cdots \oplus V_k \oplus W_1)$$
$$\cong \mathrm{Hom}_G(V, V_1) \oplus \cdots \oplus \mathrm{Hom}_G(V, V_k) \oplus \mathrm{Hom}_G(V, W_1).$$

Now $\mathrm{Dim}\,\mathrm{Hom}_G(V, V_i) = 1$ for each $i = 1, \ldots, k$ by Corollary 3.18, and $\mathrm{Dim}\,\mathrm{Hom}_G(V, W_1) = 0$ by Corollary 3.17. $\square$

5.5 Hints and Additional Comments

Exercise 5.5 One such map is $\phi_0 \colon \mathcal{P}_1 \to \mathcal{P}_{(1,1,0)}$ that is defined on the basis vectors by

$$\mathbf{x} \mapsto \mathbf{yz}, \quad \mathbf{y} \mapsto \mathbf{xz}, \quad \text{and} \quad \mathbf{z} \mapsto \mathbf{xy}.$$

Another map is $\phi_1 \colon \mathcal{P}_1 \to \mathcal{P}_{(1,1,0)}$, defined on the basis vectors by

$$\mathbf{x} \mapsto \mathbf{xy} + \mathbf{xz}, \quad \mathbf{y} \mapsto \mathbf{xy} + \mathbf{yz}, \quad \text{and} \quad \mathbf{z} \mapsto \mathbf{xz} + \mathbf{yz}.$$

Exercise 5.6 It is sufficient to just check the two-cycles since they generate S_n.

Exercise 5.8 We will work out one example in detail. We have

$$\mathcal{P}_4(x, y, z) = \mathcal{P}_{(4,0,0)} \oplus \mathcal{P}_{(3,1,0)} \oplus \mathcal{P}_{(2,2,0)} \oplus \mathcal{P}_{(2,1,1)}.$$

Decomposing each of the above summands:

$$\mathcal{P}_{(4,0,0)} = \langle\!\langle \mathbf{x}^4, \mathbf{y}^4, \mathbf{z}^4 \rangle\!\rangle \cong \mathcal{I}_3 \oplus \mathcal{W}_3 \cong \mathcal{P}_{(1,0,0)},$$

$$\mathcal{P}_{(3,1,0)} = \langle\!\langle \mathbf{x}^3\mathbf{y}, \mathbf{x}^3\mathbf{z}, \ldots, \mathbf{y}^3\mathbf{z} \rangle\!\rangle \cong \mathcal{I}_3 \oplus 2\mathcal{W}_3 \oplus \mathcal{A}_3 \cong \mathcal{P}_{(2,1,0)},$$

$$\mathcal{P}_{(2,2,0)} = \langle\!\langle \mathbf{x}^2\mathbf{y}^2, \mathbf{y}^2\mathbf{z}^2, \mathbf{x}^2\mathbf{z}^2 \rangle\!\rangle \cong \mathcal{I}_3 \oplus \mathcal{W}_3 \cong \mathcal{P}_{(1,1,0)},$$

$$\mathcal{P}_{(2,1,1)} = \langle\!\langle \mathbf{x}^2\mathbf{yz}, \mathbf{xy}^2\mathbf{z}, \mathbf{xyz}^2 \rangle\!\rangle \cong \mathcal{I}_3 \oplus \mathcal{W}_3 \cong \mathcal{P}_{(1,1,0)}.$$

Why is the decomposition of $\mathcal{P}_{(3,1,0)}$ different from the others, and why are these other decompositions all the same? The relevant observation is that it doesn't matter how you permute, for example, the variables appearing to the power 0 in $\mathcal{P}_{(4,0,0)}$, or how you permute the variables appearing to the power 1 in $\mathcal{P}_{(2,1,1)}$. We will see this again when we talk about Young permutation modules in Chap. 10.

Exercise 5.11 The Vandermonde determinant, denoted $\Delta(x, y, z)$, is given by

$$\Delta(x, y, z) := \begin{vmatrix} 1 & 1 & 1 \\ x & y & z \\ x^2 & y^2 & z^2 \end{vmatrix}.$$

It is a basic result from linear algebra that interchanging any two columns, which is just a transposition of the variables, changes the sign of the determinant.

Exercise 5.12 The polynomial

$$\triangle(\mathbf{x}) = \prod_{1 \le i < j \le n} (x_j - x_i)$$

has $n - 1$ factors of the form "$(x_n - *)$";

$$(x_n - x_1), (x_n - x_2), \ldots, (x_n - x_{n-1}),$$

$n - 2$ factors of the form "$(x_{n-1} - *)$";

$$(x_{n-1} - x_1), (x_{n-1} - x_2), \ldots, (x_{n-1} - x_{n-2}),$$

etc. Thus there are $(n - 1) + (n - 2) + \cdots + 2 + 1 = n(n - 1)/2$ factors in $\triangle(\mathbf{x})$. When $\triangle(\mathbf{x})$ is expanded, each term is a monomial that has the same type. That is, each term is a permutation of the leading term $x_2 x_3^2 \cdots x_n^{n-1}$.

If $P(\mathbf{x})$ is an alternating polynomial, then $(i, j).P(\mathbf{x}) = -P(\mathbf{x})$ for each transposition (i, j), and it follows that $P(\mathbf{x}) = 0$ if any of the variables $x_1, \ldots, x_n$ are repeated.

Define $Q_n(x_n) := P(x_1, \ldots, x_{n-1}, x_n)$. That is, consider Q_n as a polynomial in x_n alone. By the previous paragraph,

$$Q_n(x_1) = P(x_1, \ldots, x_{n-1}, x_1) = 0,$$

and so by the Fundamental Theorem of Algebra, $(x_n - x_1)$ is a factor of $Q(x_n) = P(\mathbf{x})$.[1] Repeating this reasoning, we have that each $(x_j - x_i)$ is a factor of $P(\mathbf{x})$, and consequently so is their product. It follows that $P(\mathbf{x}) = R(\mathbf{x})\triangle(\mathbf{x})$ for some quotient polynomial $R(\mathbf{x})$ that must be symmetric.

Since the Vandermonde determinant,

$$V(\mathbf{x}) = \begin{vmatrix} 1 & 1 & \cdots & 1 \\ x_1 & x_2 & & x_n \\ x_1^2 & x_2^2 & \cdots & x_n^2 \\ \vdots & \vdots & & \vdots \\ x_1^{n-1} & x_2^{n-1} & \cdots & x_n^{n-1} \end{vmatrix},$$

is alternating, it is divisible by $\triangle(\mathbf{x})$. Since each term in $V(\mathbf{x})$ is obtained by multiplying one entry from each row, $V(\mathbf{x})$ is homogeneous of degree $0 + 1 + 2 + \cdots + (n - 1) = n(n - 1)/2$. Hence $V(\mathbf{x}) = C\triangle(\mathbf{x})$ for some constant C. Comparing coefficients of the leading terms, $x_2 x_3^2 \cdots x_n^{n-1}$, we see that $C = 1$, and therefore $V(\mathbf{x}) = \triangle(\mathbf{x})$, justifying the notation in the previous exercise.

---

[1] This is sometimes referred to as the "Factor Theorem" in elementary algebra textbooks.

Exercise 5.15 Let U and W be invariant subspaces and let $\phi : V \to U \oplus W$ be a G-map. By Exercise 3.12, the projection maps $P_U : U \oplus W \to U$ and $P_W : U \oplus W \to W$ are G-maps. Since compositions and linear combinations of G-maps are G-maps, we have

$$P_U \circ \phi + P_W \circ \phi \in \operatorname{Hom}_G(V, U) \oplus \operatorname{Hom}_G(V, W).$$

Conversely, if $\phi_U \in \operatorname{Hom}_G(V, U)$ and $\phi_W \in \operatorname{Hom}_G(V, W)$, then $\phi_U + \phi_V \in \operatorname{Hom}_G(V, U \oplus W)$. Hence,

$$\operatorname{Hom}_G(V, U \oplus W) \cong \operatorname{Hom}_G(V, U) \oplus \operatorname{Hom}_G(V, W).$$

The Group Algebra

6

The group algebra is an important construction that assigns an algebra to a group. In this chapter we present two equivalent versions of the group algebra, and derive some results for later consumption.

Remark* 6.1 For our purposes, an *associative algebra* is a vector space with the additional operation of associative multiplication of vectors. Two examples that you should be familiar with are $\mathcal{P}(x_1, \ldots, x_n)$, the algebra of polynomials in n variables or indeterminants, and $\mathrm{Mat}_{k,k}(\mathbb{F})$, the algebra of $k \times k$ matrices with entries from a field $\mathbb{F}$, both with the usual operations. All algebras are assumed to be associative, and to contain a multiplicative identity unless explicitly stated otherwise. We could also define an algebra as a *ring* with the additional operation of scalar multiplication. Curious readers can look up the definition of a ring.

6.1   Version One

Given a field such as $\mathbb{C}$ and a finite group G, we can construct a $\mathbb{C}$-vector space, called the *group algebra* and denoted $\mathbb{C}[G]$, by declaring that the elements of G are a basis, and then formally taking finite linear combinations over $\mathbb{C}$. For example, $3(\mathbf{1, 2}) + \pi\,(\mathbf{1, 3, 2}) - \frac{2}{3}(\mathbf{1, 3}) \in \mathbb{C}[S_3]$. We will write the group elements in boldface when we are considering them as vectors. We give $\mathbb{C}[G]$ an algebra structure by declaring that the group multiplication (which is associative) be distributive over vector addition, and we will denote vector multiplication by $\times$ when needed for clarity. For example, in $\mathbb{C}[S_3]$,

$$[(\mathbf{1}) + 3(\mathbf{1, 2, 3})] \times [(\mathbf{1, 2}) - (\mathbf{1, 3})] = (\mathbf{1, 2}) + 2(\mathbf{1, 3}) - 3(\mathbf{2, 3}).$$

Since $\mathbb{C}[G]$ is a vector space, we can define a representation of G on $\mathbb{C}[G]$, called the *left regular representation* and denoted L, by having an element $\sigma \in G$

© The Author(s), under exclusive license to Springer Nature Switzerland AG 2022
R. M. Howe, *An Invitation to Representation Theory*, SUMS Readings,
https://doi.org/10.1007/978-3-030-98025-2_6

act linearly on $\mathbf{v} = r_1\mathbf{g_1} + \cdots + r_n\mathbf{g_n} \in \mathbb{C}[G]$ (for $g_i \in G$ and $r_i \in \mathbb{C}$):

$$L(\sigma)\mathbf{v} = L(\sigma)(r_1\mathbf{g_1} + \ldots + r_n\mathbf{g_n}) = r_1(\sigma\mathbf{g_1}) + \ldots + r_n(\sigma\mathbf{g_n}).$$

To be thorough, we record that there is also a *right regular representation*, denoted R:

$$R(\sigma)\mathbf{v} = R(\sigma)(r_1\mathbf{g_1} + \ldots + r_n\mathbf{g_n}) = r_1(\mathbf{g_1}\sigma^{-1}) + \ldots + r_n(\mathbf{g_n}\sigma^{-1}).$$

Exercise 6.2 Practice some computations with the group algebra $\mathbb{C}[S_3]$. Verify that the left and right regular representations are, in fact, representations.

Remark* 6.3 If a group G acts on a vector space V, then the G-action extends by linearity to an action of the group algebra $\mathbb{C}[G]$ on V. That is, if (ρ, V) is a representation of G on a vector space V, and if $a = \sum_{g \in G} r_g \mathbf{g} \in \mathbb{C}[G]$ for some scalars r_g, then the action of $\mathbb{C}[G]$ on V is given by $a.\mathbf{v} = \sum_{g \in G} r_g \rho(g)\mathbf{v}$. Thus the representations of $\mathbb{C}[G]$ correspond exactly to the representations of G and vice versa. Whether we wish to study the representations of the group or of its group algebra is largely a matter of taste or convenience. We obtain the sharpest results when we take the complex numbers $\mathbb{C}$ as the field of scalars since $\mathbb{C}$ is algebraically closed and of characteristic zero.

Remark 6.4 Every irreducible representation of finite group G appears in the group algebra $\mathbb{C}[G]$. Sketch of proof: If W is an irreducible representation of G and if $\phi\colon \mathbb{C}[G] \to W$ is a non-zero G-map, then $\mathbb{C}[G] \cong W \oplus \mathrm{Ker}\,\phi$. A more detailed proof requires the notion of quotient spaces, which we review in Sect. 9.2 (but may be known to some readers). See Exercise 9.8.

Definition 6.5 If an algebra A acts on a vector space V, then we say that V has the structure of an *A-module*. If A acts on V on the left (right), we say V is a left (right) A-module. The term "module" arises in other mathematical contexts as we have noted previously.

Example 6.6 The group algebra $\mathbb{C}[S_n]$ is an example of a $\mathbb{C}[S_n]$-bimodule, with the left and right actions given by the left and right regular representations (which are required to commute with each other). The space $\mathcal{P}$ of polynomials in n variables is a left $\mathbb{C}[S_n]$-module. How could we define some sort of right action of S_n on $\mathcal{P}$ to make $\mathcal{P}$ into a right $\mathbb{C}[S_n]$-module?

Definition 6.7 Similar to the center of a group (Exercise 1.63), we can define the *center of the group algebra*, denoted $\mathcal{Z}\mathbb{C}[G]$, as those elements in the group algebra that commute with all other elements in the group algebra;

$$\mathcal{Z}\mathbb{C}[G] := \{a \in \mathbb{C}[G] \mid ab = ba \text{ for all } b \in \mathbb{C}[G]\}.$$

Exercise* 6.8 Determine $\mathcal{Z}\mathbb{C}[S_3]$. Hint: It is sufficient to determine an arbitrary vector in $\mathbb{C}[S_3]$ that commutes with the transpositions. Generalize this to determine $\mathcal{Z}\mathbb{C}[S_n]$ for an arbitrary n.

Exercise 6.9 Given an algebra A, how would we define a sub-algebra of A? Show that $\mathcal{Z}\mathbb{C}[G]$ is a sub-algebra of $\mathbb{C}[G]$. Hint: Apply the definitions to show that $\mathcal{Z}\mathbb{C}[G]$ is a vector sub-space of $\mathbb{C}[G]$ that is closed under vector multiplication.

A useful feature of central elements is that they "act like G-maps."

Exercise 6.10 If $z \in \mathcal{Z}\mathbb{C}[G]$ is central, show that the map $\phi_z \colon \mathbb{C}[G] \to \mathbb{C}[G]$ given by $\phi_z(v) = zv$ intertwines the left regular representation of G on $\mathbb{C}[G]$.

Remark 6.11 One early interpretation of an element in $\mathbb{C}[S_n]$ was via an action on functions, (see [R]). For example, an element such as $(1, 2) - \frac{2}{3}(1, 3)$ (called a *substitutional expression*) meant "apply this to a function $F(x, y, z)$ by permuting the variables." That is,

$$(1, 2) - \frac{2}{3}(1, 3) \text{ applied to } F(x, y, z) \text{ becomes } F(y, x, z) - \frac{2}{3}F(z, y, x).$$

Similarly, the identity

$$f(x, y) = \frac{1}{2}[f(x, y) + f(y, x)] + \frac{1}{2}[f(x, y) - f(y, x)]$$

is replace by the *substitutional equation;*

$$(1) = \frac{1}{2}[(1) + (1, 2)] + \frac{1}{2}[(1) - (1, 2)].$$

6.2 Version Two

There is another important incarnation of the group algebra for a finite group G;

$$\mathbb{C}[G] := \{a \colon G \to \mathbb{C}\},$$

the set of all complex-valued functions[1] on the group G. We have the usual operations of vector addition and scalar multiplication in function spaces (Sect. 1.6), but multiplication of vectors is given by the *convolution product*, denoted by $*$;

---

[1] This notation coincides with that used in Example 3.32.

Definition 6.12

$$[a * b](x) := \sum_{y \in G} a(xy^{-1})b(y) \quad \text{for all } a, b \in \mathbb{C}[G], x \in G.$$

Why such a goofy multiplication? One reason is because this provides an explicit algebra isomorphism between these two versions of $\mathbb{C}[G]$.

Exercise* 6.13 Define $\delta_g : G \to \mathbb{C}$ for each $g \in G$ as

$$\delta_g(x) := \begin{cases} 1, & \text{if } x = g; \\ 0, & \text{otherwise.} \end{cases}$$

Show that the set $\{\delta_g \mid g \in G\}$ is a basis for $\mathbb{C}[G]$, and therefore the dimension of $\mathbb{C}[G] = |G|$.

Exercise* 6.14 Show that $\delta_g * \delta_h = \delta_{gh}$. Consequently, the map $g \mapsto \delta_g$ (extended by linearity) is an algebra isomorphism between the two incarnations of $\mathbb{C}[G]$.

The left and right regular representations are then defined as $L(g)a(x) = a(g^{-1}x)$, and $R(g)a(x) = a(xg)$ respectively, for $a \in \mathbb{C}[G]$ and $x, g \in G$.

Exercise* 6.15 Show that the the left and right regular representations of G on $\mathbb{C}[G]$ are equivalent via the G-map $a \mapsto \check{a}$ given by $\check{a}(x) = a(x^{-1})$.

Notation 6.16 We will use the notation $\mathbb{C}[G]$ for both incarnations of the group algebra when it is clear from the context or immaterial.

How do we describe the center $\mathcal{Z}\mathbb{C}[G]$ for this version of $\mathbb{C}[G]$? By linearity, it is sufficient to consider the basis elements $\{\delta_g\}$, and determine the properties of any $a \in \mathbb{C}[G]$ so that $a * \delta_g = \delta_g * a$ for all $g \in G$. On one hand,

$$[a * \delta_g](x) = \sum_{y \in G} a(xy^{-1})\delta_g(y) = a(xg^{-1}) \quad \text{for all } x \in G$$

since each term is zero unless $g = y$, and thus the only term that survives is $a(xg^{-1})$.
On the other hand,

$$[\delta_g * a](x) = \sum_{y \in G} \delta_g(xy^{-1})a(y) = a(g^{-1}x) \quad \text{for all } x \in G.$$

Consequently, $a \in \mathcal{Z}\mathbb{C}[G]$ exactly when $a(xy) = a(yx)$ for all $x, y \in G$. Such functions are called *central functions*.

But wait, there's more! If $a(xy) = a(yx)$, then we can set $g = xy$, and the relation $a(xy) = a(yx)$ becomes $a(g) = a(ygy^{-1})$. Hence central functions are also constant on conjugacy classes.

Definition 6.17 The set of functions from G to $\mathbb{C}$ that are constant on conjugacy classes of G are called *class functions*, denoted $\mathbb{C}_{class}[G]$.

Conversely, suppose that $a \in \mathbb{C}[G]$ is a class function. Then

$$a(gh) = a(h(gh)h^{-1}) = a(hg), \quad \text{and hence } a \text{ is a central function.}$$

Compare this result with the results from Exercise 6.8.

Exercise 6.18 Show that $\mathbb{C}_{class}[G]$ is a sub-algebra of $\mathbb{C}[G]$. That is, show that $\mathbb{C}_{class}[G]$ is a vector sub-space of $\mathbb{C}[G]$ that is closed under vector multiplication. Do the same for $Z\mathbb{C}[G]$.

We have proven the following Proposition.

Proposition 6.19 *Let G be a finite group, and let $\mathbb{C}[G]$ be the group algebra with the convolution product. Let $Z\mathbb{C}[G]$ be the center of $\mathbb{C}[G]$, and let $\mathbb{C}_{class}[G]$ be the sub-algebra of class functions on G. Then $Z\mathbb{C}[G] = \mathbb{C}_{class}[G]$.*

Remark 6.20 With some extra work, version two of the group algebra makes sense when $|G| = \infty$. Again we define $\mathbb{R}[G]$ or $\mathbb{C}[G]$ as the vector space of all $\mathbb{R}$-valued or $\mathbb{C}$-valued functions on G, but the convolution product defined as a sum only makes sense if the functions on G are *finitely supported*, that is, $a(x) = 0$ for all but finitely many $x \in G$.

For some groups we can define the convolution product by

$$(f_1 * f_2)(x) = \int_G f_1(xg^{-1}) f_2(g)\, dg.$$

That is, the finite sum in Definition 6.12 is replaced by integration over the group, so there has to be some machinery in place for such integration to make sense. There also have to be some conditions on everything involved so that the integral exists, is convergent, etc. Such considerations are beyond the modest scope of this book.

6.3 Hints and Additional Comments

Remark 6.1 There are also non-associative algebras. One of the more important examples is a *Lie algebra* $\mathfrak{g}$, which is a vector space with the product of vectors given by the *Lie bracket*, $[\cdot, \cdot]$, that satisfies the following properties:

- Bilinearity:

$$[r\mathbf{u} + s\mathbf{v}, \mathbf{w}] = r[\mathbf{u}, \mathbf{w}] + s[\mathbf{v}, \mathbf{w}], \quad \text{and} \quad [\mathbf{w}, r\mathbf{u} + s\mathbf{v}] = r[\mathbf{w}, \mathbf{u}] + s[\mathbf{w}, \mathbf{v}],$$

  for all scalars r, s, and all $\mathbf{u}, \mathbf{v}, \mathbf{w} \in \mathfrak{g}$.
- The Jacobi identity:

$$[\mathbf{u}, [\mathbf{v}, \mathbf{w}]] + [\mathbf{w}, [\mathbf{u}, \mathbf{v}]] + [\mathbf{v}, [\mathbf{w}, \mathbf{u}]] = \mathbf{0}, \text{ for all } \mathbf{u}, \mathbf{v}, \mathbf{w} \in \mathfrak{g}.$$

- Alternativity:

$$[\mathbf{u}, \mathbf{u}] = \mathbf{0} \quad \text{for all } \mathbf{u} \in \mathfrak{g}.$$

Any associative algebra A, such as $\mathrm{Mat}_{k,k}(\mathbb{R})$, can be given a Lie algebra structure by defining the bracket product to be the *commutator product*, defined by

$$[\mathbf{u}, \mathbf{v}] := \mathbf{u}\mathbf{v} - \mathbf{v}\mathbf{u} \quad \text{for all } \mathbf{u}, \mathbf{v} \in A.$$

You should check that this satisfies the above properties of a Lie algebra. Of course, if the associative algebra is commutative, then the commutator product is identically zero.

Lie algebras (and the related Lie groups) are named after Sophus Lie, a Norwegian mathematician. Therefore, the correct pronunciation of Lie sounds like "Lee." A Lie algebra that you should already be familiar with is the vector space $\mathbb{R}^3$, with the Lie bracket given by $[\mathbf{u}, \mathbf{v}] = \mathbf{u} \times \mathbf{v}$, the vector cross product.

Remark 6.3 A representation of an algebra A on a vector space V is an algebra homomorphism from A to the algebra $\mathrm{End}(V)$, the algebra of endomorphisms on V, which can be realized (after choosing an ordered basis for V) as the algebra of $k \times k$ matrices, where $k = \mathrm{Dim}(V)$.

Exercise 6.8 Let

$$\mathbf{z} = r_1(\mathbf{1}) + r_2(\mathbf{1}, \mathbf{2}) + r_3(\mathbf{1}, \mathbf{3}) + r_4(\mathbf{2}, \mathbf{3}) + r_5(\mathbf{1}, \mathbf{2}, \mathbf{3}) + r_6(\mathbf{1}, \mathbf{3}, \mathbf{2})$$

be an arbitrary vector in $\mathcal{Z}\mathbb{C}[S_3]$. Then $(\mathbf{1}, \mathbf{2})\mathbf{z} = \mathbf{z}(\mathbf{1}, \mathbf{2})$ implies that $r_3 = r_4$ and $r_5 = r_6$. Similar computations with the other transpositions (which generate S_3) gives $r_2 = r_3 = r_4$ and $r_5 = r_6$. That is, the coefficients are constant for each cycle type, which is the same as saying that the coefficients are constant on each conjugacy class. It follows that $\mathcal{Z}\mathbb{C}[S_3]$ has

$$\{(\mathbf{1}), \ (\mathbf{1}, \mathbf{2}) + (\mathbf{1}, \mathbf{3}) + (\mathbf{2}, \mathbf{3}), \ (\mathbf{1}, \mathbf{2}, \mathbf{3}) + (\mathbf{1}, \mathbf{3}, \mathbf{2})\}$$

as a basis, and thus has dimension 3, the number of conjugacy classes in S_3.

More generally, if $\mathbf{z} = \sum_{g \in G} r_g \mathbf{g} \in \mathcal{Z}\mathbb{C}[G]$ for some scalars $r_g \in \mathbb{C}$, and for each $h \in G$, we have

$$\mathbf{z} = h\mathbf{z}h^{-1} = h\left[\sum_{g \in G} r_g \mathbf{g}\right]h^{-1} = \sum_{g \in G} r_g h\mathbf{g}h^{-1} = \sum_{g \in G} r_{h^{-1}gh}\mathbf{g}.$$

Since $\{\mathbf{g}\}_{g \in G}$ is a basis for $\mathbb{C}[G]$, we can equate coefficients to obtain $r_g = r_{h^{-1}gh}$ for all $g \in G$ and any $h \in G$. In other words, the coefficients are constant on the conjugacy classes, and thus the dimension of $\mathcal{Z}\mathbb{C}[G]$ is equal to the number of conjugacy classes in G. It is routine to check that the converse is also true; if the coefficients of a vector $a \in \mathbb{C}[G]$ are constant on the conjugacy classes, then a is central.

Exercise 6.13 Let $f \in \mathbb{C}[G]$. Then $f(x) = \sum_{g_i \in G} f(x)\delta_{g_i}(x)$, so the set $\{\delta_{g_i} \mid g_i \in G\}$ spans $\mathbb{C}[G]$. It is routine to check that this set is also linearly independent, since if

$$\sum_{g_i \in G} a_{g_i}\delta_{g_i}(x) = 0 \text{ for all } x \in G,$$

set $x = g_j$ to obtain $a_{g_j} = 0$ for each $g_j \in G$.

Exercise 6.14 For every $g, h,$ and x in G we have

$$(\delta_g * \delta_h)(x) = \sum_{y \in G} \delta_g(xy^{-1})\delta_h(y).$$

The only term that survives, and in which case it is equal to 1, is when $y = h$ and $g = xy^{-1} = xh^{-1}$, so that $x = gh$. But this is the definition of $\delta_{gh}(x)$.

An *algebra homomorphism* is a linear map between vector spaces that also preserves the operation of vector multiplication. It is an *algebra isomorphism* if it is also a bijection.

Exercise 6.15 It helps to first clarify the problem. Let $\phi \colon \mathbb{C}[G] \to \mathbb{C}[G]$ be given by $\phi(a) = \check{a}$, where $\check{a}(x) = a(x^{-1})$ for $a \in \mathbb{C}[G]$ and $x \in G$. Then for any $g \in G$,

$$\begin{aligned}
[L(g)\phi(a)](x) &= [\phi(a)](g^{-1}x) \\
&= \check{a}(g^{-1}x) \\
&= a(x^{-1}g) \\
&= [R(g)a](x^{-1}) \\
&= [R(g)\check{a}](x) \\
&= [R(g)\phi(a)](x).
\end{aligned}$$

Thus ϕ intertwines the left and right regular representations of G on $\mathbb{C}[G]$.

The Irreducible Representations of S_n: Characters

7

In the next two chapters we determine, label, and construct all of the distinct irreducible representations of S_n. In this chapter we define a special kind of function on the group, called the character of a representation, and then show that two representations are equivalent exactly when their characters are equal. The characters of the irreducible representations are in one-to-one correspondence with the conjugacy classes in S_n, which in turn are in one-to-one correspondence with the partitions of n, so we can use partitions of n to label the irreducible representations of S_n. Group characters are a useful tool in the study of group representations, and any book on representation theory should include their exposition, but they are not essential to the more "constructive" approach that is our focus. In the next chapter we will actually construct these irreducible representations in the group algebra $\mathbb{C}[S_n]$, and use these representations to obtain irreducible representations in other vector spaces.

The discussion in this chapter is adapted from a variety of sources. See [FH] Lecture 2, [St] Chapter 2 and [Si] Chapter III. The proofs and other details in this chapter are necessarily more technical than those previously encountered. Readers new to the subject may wish to focus on the results, and digest the details of the proofs later. Observant readers will note that most of these results apply to any finite group.

7.1 Characters and Class Functions

Characters were introduced by G. Frobenius in the 1890s, and were studied before their connection to representation theory was fully understood (see [C]). Characters are a powerful tool for the study of groups and group representations since it is often easier to determine the character of a representation than to find an actual realization on a vector space. Much of the literature of representation theory, especially for finite groups, is concerned with characters.

© The Author(s), under exclusive license to Springer Nature Switzerland AG 2022
R. M. Howe, *An Invitation to Representation Theory*, SUMS Readings,
https://doi.org/10.1007/978-3-030-98025-2_7

Let A be a square matrix, and recall tr(A), the *trace* of A, which is the sum of the diagonal entries of A. Equivalently, if V is a vector space and $T : V \to V$ is linear, then tr(T) is the sum of the diagonal entries in the matrix realization of T. It is worthwhile to recall a few of the properties of the trace for $k \times k$ matrices A and B:

(1) tr$(A + B)$ = tr(A) + tr(B).
(2) tr(AB) = tr(BA).
(2') If A is invertible, such as a change-of-basis matrix, then

$$\text{tr}[A(BA^{-1})] = \text{tr}[(BA^{-1})A] = \text{tr}(B).$$

Hence the trace of a linear operator is well defined, that is, independent of its matrix realization.

Remark 7.1 When a square matrix A is diagonalizable, or when we are working in a complex vector space (our default situation), then tr(A) is the sum of the eigenvalues of A, or equivalently, the sum of the eigenvalues of a linear operator with A as a matrix realization. This follows from that property that the *characteristic equation*, $\det(xI - A) = 0$, has n solutions $\{\alpha_i\}$ (counting multiplicities) in $\mathbb{C}$. In this case there is an invertible matrix P (a change of basis matrix) such that $PAP^{-1} = J$, where J is in *Jordan normal form*:

$$J = \begin{pmatrix} \alpha_1 & 1 & 0 & & \cdots & 0 \\ 0 & \alpha_2 & 1 & 0 & \cdots & 0 \\ 0 & 0 & \alpha_3 & 1 & \cdots & 0 \\ \vdots & & 0 & \ddots & & \vdots \\ 0 & \cdots & & 0 & \alpha_{n-1} & 1 \\ 0 & & \cdots & & 0 & \alpha_n \end{pmatrix}.$$

A detailed exposition can be found in most any serious linear algebra text, for example [L2].

Exercise* 7.2 Show that if V is a finite-dimensional vector space with subspace U, and if P_U is the projection onto U, then tr(P_U) = Dim U.
Let $\phi : V \to V^G$ be the projection from Exercise 2.28. Conclude that tr(ϕ) = Dim V^G.

Definition 7.3 Let (ρ, V) be a finite-dimensional representation of a group G. Define the *character* of V, $\chi_V : G \to \mathbb{C}$, given by

$$\chi_V(g) := \text{tr}[\rho(g)] = \text{tr}(\tilde{g}).$$

Characters of irreducible representations are called *irreducible characters*.

Proposition 7.4 *Several important properties of characters follow immediately from the properties of the trace:*

(1) If V is a representation of G, and if $V = V_1 \oplus V_2 \oplus \cdots \oplus V_k$, where each V_i is a G-invariant subspace of V, then $\chi_V = \chi_{V_1} + \chi_{V_2} + \cdots + \chi_{V_k}$.

(2) Characters are constant on conjugacy classes; $\chi_V(hgh^{-1}) = \chi_V(g)$, and hence characters are class functions.

Recall from Definition 6.17 that the set of all functions from a group G to the field $\mathbb{C}$ that are constant on the conjugacy classes of G is called the space of *class functions* on G, denoted $\mathbb{C}_{class}[G]$. By Proposition 6.19, the space of class functions is equal to $\mathcal{Z}\mathbb{C}[G]$, the center of $\mathbb{C}[G]$.

Exercise* 7.5 Let $[g]$ denote the conjugacy class of g in a finite group G. Define $\delta_{[g]} : G \to \mathbb{C}$ by

$$\delta_{[g]}(h) := \begin{cases} 1, & \text{if } h \in [g]; \\ 0, & \text{otherwise.} \end{cases}$$

Show that the set $\{\delta_{[g]} \mid g \in G\}$ is a basis for $\mathbb{C}_{class}[G]$ regarded as a $\mathbb{C}$-vector space, and consequently the dimension of $\mathbb{C}_{class}[G]$ is equal to the number of conjugacy classes in G. Compare this result with the results of Exercise 6.8 and Proposition 6.19.

Exercise 7.6 Choose one element from each conjugacy class of S_3, and write out the explicit matrix realizations for the action of S_3 on, say $\mathcal{P}_1(x, y, z)$, or equivalently, on the permutation representation from Exercise 1.22. Determine the character of each such element. Generalize this result to show that, for a permutation representation V of S_n, the character $\chi_V(\sigma)$ is the number of basis vectors fixed by σ.

Exercise 7.7 Show that if V is a representation of a group G with identity element e, then $\chi_V(e) = \text{Dim } V$.

Exercise 7.8 Let (ρ, V) be a representation of G, and let (ρ^*, V^*) be the contragradient representation from Sect. 3.4. Show that $\chi_{V^*}(g) = \chi_V(g^{-1})$.

Proposition 7.9 *If (ρ, V) and (π, W) are equivalent representations of a group G, then $\chi_V(g) = \chi_W(g)$ for all $g \in G$.*

Proof If (ρ, V) and (π, W) are equivalent, then there is an invertible G-map $\phi \colon V \to W$ such that $\phi[\rho(g)\mathbf{v}] = \pi(g)\phi(\mathbf{v})$ for all $\mathbf{v} \in V$ and all $g \in G$. Then, as linear maps,

$$\phi\rho(g) = \pi(g)\phi \Rightarrow \rho(g) = \phi^{-1}\pi(g)\phi$$
$$\Rightarrow \operatorname{tr}[\rho(g)] = \operatorname{tr}[\phi^{-1}\pi(g)\phi] = \operatorname{tr}[\pi(g)]$$
$$\Rightarrow \chi_V(g) = \chi_W(g).$$

$\square$

Proposition 7.10 *The converse of Proposition 7.9 is also true; if two representations of a group have the same character, then they are equivalent.*

Proof This follows from the fact that the irreducible characters are a basis for the vector space $\mathbb{C}_{class}[G]$, which is proven in Proposition 7.18 below. Consequently, suppose that V and W are two representation of G, and let χ_i be the irreducible character for each distinct irreducible representation V_i. Then if

$$\chi_V = \chi_W = r_1\chi_1 + \cdots + r_k\chi_k, \quad \text{it follow that} \quad V \cong r_1V_1 \oplus \cdots \oplus r_kV_k \cong W$$

since the scalars r_i are uniquely determined. $\square$

7.2   Characters of S_3

Towards our goal of generating examples, we can now work out the characters for the irreducible representations of S_3 which we place in a *character table*. The top row is a representative from each conjugacy class, and the number in brackets is the number of permutations in each conjugacy class. Determining the characters for the one-dimensional trivial and alternating representations is easy since their eigenvalues are obvious. To determine the character of the irreducible representation $\mathcal{W}$, we use the character table for $\mathcal{P}_1 \cong \mathcal{I} \oplus \mathcal{W}$, Exercise 7.6, and property (1) in Proposition 7.4, subtracting the character for $\mathcal{I}$ from that of $\mathcal{P}_1$;

S_3	(1)	(1, 2)	(1, 2, 3)
Permutation $\simeq \mathcal{P}_1$	3	1	0
$\mathcal{I}$	1	1	1
$\mathcal{W}$	2	0	-1

The table for the irreducible characters of S_3. It is not a coincidence that this table is square.

S_3	[1] (1)	[3] (1, 2)	[2] (1, 2, 3)
Trivial $\mathcal{I}$	1	1	1
Alternating $\mathcal{A}$	1	-1	1
Standard $\mathcal{W}$	2	0	-1

Exercise 7.11 Choose a basis for the representation $\mathcal{W}_3$ of S_3 in $\mathcal{P}_1(x, y, z)$, and determine the matrix realization of one element from each conjugacy class of S_3. Verify the values of $\chi_{\mathcal{W}}(g)$ in the above table. It is instructive to repeat this using a different basis for $\mathcal{W}$.

Exercise* 7.12 Compute the character for the representation of S_3 on $\mathcal{P}_{(2,1,0)}$ $(x, y, z) \cong \mathcal{I} \oplus 2\mathcal{W} \oplus \mathcal{A}$, and verify that it satisfies the addition property for characters.

Exercise* 7.13 Check that the irreducible characters of S_3 are linearly independent. That is, if

$$r\chi_{\mathcal{I}}(\sigma) + s\chi_{\mathcal{W}}(\sigma) + t\chi_{\mathcal{A}}(\sigma) = 0 \text{ for scalars } r, s, t, \text{ and for all } \sigma \in S_3,$$

then $r = s = t = 0$.

Remark 7.14 It turns out that the characters for the representations of the symmetric group are always integer-valued. This is not true in general, even for an uncomplicated finite group such as the alternating group A_3. The proof requires sturdier algebraic tools than we can develop in this brief text.

7.3   Orthogonality of Characters, Bases

In this section we show that the irreducible characters are an orthonormal basis for $\mathbb{C}_{class}[G]$, the class functions on a group G. Along the way we derive some auxiliary results regarding matrix entries.

Recall that if U and V are $\mathbb{F}$-vector spaces of dimension k and n respectively, then, upon assigning bases to U and V, we can identify $\text{Hom}(U, V)$ with $\text{Mat}_{n,k}(\mathbb{F})$, the space of $n \times k$ matrices with entries in $\mathbb{F}$. If we let $E_{i,j}$ denote the elementary matrix with i, j-entry equal to one and with all other entries equal to zero, then the set $\{E_{i,j} \mid i = 1 \ldots n, j = 1 \ldots k\}$ is a basis for $\text{Mat}_{n,k}(\mathbb{F})$. By Exercise 3.6, $\text{Hom}_G(U, V)$, the set of G-maps from U to V, is a subspace of $\text{Hom}(U, V)$.

Definition 7.15 As in Definition 1.87, given a representation π of a group G on an $\mathbb{C}$-vector space V, with matrix realization $\tilde{\pi}(g)$ for $g \in G$, the entries $\tilde{\pi}_{i,j}(g) \in \mathbb{C}$ are called the *matrix entries* of the representation, and are functions in $\mathbb{C}[G]$, the group algebra of functions from G to $\mathbb{C}$. Of course, the value of the entries in any matrix realization depend on the basis that is chosen for V.

For example, for the representation π in Exercise 2.34, and for $(2, 3) \in S_3$, we have

$$\tilde{\pi}(2, 3) = \begin{pmatrix} \frac{1}{2} & \frac{\sqrt{3}}{2} \\ \frac{\sqrt{3}}{2} & -\frac{1}{2} \end{pmatrix},$$

with $\tilde{\pi}_{1,1}(2, 3) = \frac{1}{2}$ and $\tilde{\pi}_{1,2}(2, 3) = \frac{\sqrt{3}}{2}$.

Such matrix entries have remarkable and useful properties.

Proposition 7.16 *Let (ρ, U) and (π, V) be inequivalent irreducible unitary representations of a finite group G on finite-dimensional complex vector spaces. Then the matrix entries for the representations π and ρ of G are orthogonal with respect to the inner product on $\mathbb{C}[G]$ (Sect. 3.3) given by*

$$\langle \alpha, \beta \rangle = \frac{1}{|G|} \sum_{g \in G} \alpha(g)\overline{\beta(g)}, \quad \alpha, \beta \in \mathbb{C}[G].$$

More explicitly:

(1) $\langle \tilde{\pi}_{i,j}, \tilde{\rho}_{\ell,k} \rangle = 0$ *for all i, j, k, ℓ whenever $\rho \not\cong \pi$.*
   Moreover,

(2) $\langle \tilde{\pi}_{i,j}, \tilde{\pi}_{k,\ell} \rangle = \begin{cases} 1/(degree\ \pi) & if\ i = k\ and\ j = \ell; \\ 0 & otherwise. \end{cases}$

Proof The proof utilizes the "averaging over G" trick, and requires complex scalars.

(1) Let $S: U \to V$ be any linear map, and define $\mathbf{S}: U \to V$ by averaging S over G;

$$\mathbf{S} := \frac{1}{|G|} \sum_{g \in G} \pi(g)S\rho(g^{-1}).$$

(2) The map **S** intertwines (ρ, U) and (π, V) since for any $h \in G$:

$$\pi(h)\mathbf{S} = \pi(h)\frac{1}{|G|}\sum_{g\in G}\pi(g)S\rho(g^{-1})$$

$$= \frac{1}{|G|}\sum_{g\in G}\pi(h)\pi(g)S\rho(g^{-1})$$

$$= \frac{1}{|G|}\sum_{g\in G}\pi(hg)S\rho(g^{-1})$$

$$(\text{setting } hg = x) = \frac{1}{|G|}\sum_{x\in G}\pi(x)S\rho(x^{-1}h)$$

$$= \frac{1}{|G|}\sum_{x\in G}\pi(x)S\rho(x^{-1})\rho(h)$$

$$= \left[\frac{1}{|G|}\sum_{x\in G}\pi(x)S\rho(x^{-1})\right]\rho(h)$$

$$= \mathbf{S}\,\rho(h).$$

Here we have used the fact that summing over all $g \in G$ is the same as summing over $x = hg$ for a fixed h and all $g \in G$.

(3) Since (ρ, U) and (π, V) are both irreducible complex representations, we can apply the strong version of Schur's Lemma 3.16 to obtain:

 (a) The map **S** is identically zero since the representations (ρ, U) and (π, V) are inequivalent.

 (b) If $\rho = \pi$ and $U = V$ (with dimension d), then the map **S** is a scalar multiple c of the identity map $\mathbb{I}$. Note that Complex scalars are required here.

(4) For case (b), we have

$$\operatorname{tr}\mathbf{S} = \operatorname{tr}\left[\frac{1}{|G|}\sum_{g\in G}\pi(g)S\pi(g^{-1})\right]$$

$$= \frac{1}{|G|}\sum_{g\in G}\operatorname{tr}\left[\pi(g)S\pi(g^{-1})\right]$$

$$= \frac{1}{|G|}\sum_{g\in G}[\operatorname{tr} S]$$

$$= \operatorname{tr} S.$$

Thus,

$$\text{tr } \mathbf{S} = \text{tr } c\mathbb{I} = cd = \text{tr } S, \text{ so the constant } c = \frac{1}{d} \text{ tr } S.$$

(5) Choose bases for U and V. Using our usual notation, we let $\widetilde{\mathbf{S}}$ denote the matrix realization of the linear map $\mathbf{S}$ with entries $\widetilde{\mathbf{S}}_{i,\ell}$, and similarly for the other linear maps involved. We translate the definition of the map $\mathbf{S}$ into the language of matrix multiplication, and obtain an identity for the matrix entries;

$$\widetilde{\mathbf{S}}_{i,\ell} = \frac{1}{|G|} \sum_{g \in G} \sum_{j,k} \widetilde{\pi(g)}_{i,j} \widetilde{S}_{j,k} \widetilde{\rho(g^{-1})}_{k,\ell}.$$

(6) We record the following observation here in order to streamline the ensuing arguments. If E_{jk} is a standard basis matrix, and if A and B are matrices with the appropriate dimensions, then

$$(A E_{jk} B)_{i\ell} = A_{ij} B_{k\ell}.$$

Indeed, from linear algebra we know that

$$(A E_{jk} B)_{i\ell} = \sum_{r,t} A_{ir} (E_{jk})_{rt} B_{t\ell},$$

but $(E_{jk})_{rt} = 0$ unless $r = j$ and $t = k$, in which case $(E_{jk})_{jk} = 1$. In case there is any confusion, the expression $(E_{jk})_{rt}$ means the r, t-entry in the basis matrix E_{jk}.

Exercise 7.17 Work through some simple examples that illustrates this result.

(7) When we let S be a linear map whose matrix realization is a basis matrix E_{jk}, the previous observation yields

$$(\mathbf{E_{jk}})_{i,\ell} = \frac{1}{|G|} \sum_{g \in G} \left[\widetilde{\pi(g)} E_{jk} \widetilde{\rho(g^{-1})} \right]_{i\ell} = \frac{1}{|G|} \sum_{g \in G} \widetilde{\pi(g)}_{i,j} \widetilde{\rho(g^{-1})}_{k,\ell}.$$

(8) We can always choose bases for U and V that are orthonormal with respect to a G-invariant inner product, in which case the representations π and ρ can be realized by unitary matrices (Exercise 3.43 and Remark 3.44). Since $A^{-1} = \overline{A}^T$ when A is unitary, we have the following identity for all i, j, k, ℓ when ρ and π are inequivalent:

$$0 = (\mathbf{E_{jk}})_{i,\ell} = \frac{1}{|G|} \sum_{g \in G} \widetilde{\pi(g)}_{i,j} \overline{\widetilde{\rho(g)}}_{\ell,k} = \langle \widetilde{\pi}_{i,j}, \widetilde{\rho}_{\ell,k} \rangle.$$

Thus, the matrix entries for inequivalent irreducible unitary representations are orthogonal.

(9) We apply the same reasoning when $\rho = \pi$ and $U = V$. In this case $\mathbf{S} = c\mathbb{I}$ for some constant c, and from part (4) we have

$$\operatorname{tr} \mathbf{E_{jk}} = \frac{\operatorname{tr}(E_{jk})}{d}\mathbb{I}.$$

Since $\operatorname{tr} E_{jk} = 0$ if $j \neq k$, our previous computations show that

$$(\mathbf{E_{jk}})_{i\ell} = \langle \tilde{\pi}_{i,j}, \tilde{\pi}_{\ell,k} \rangle = 0 \text{ when } j \neq k.$$

Since $(\mathbf{E_{jk}})_{i\ell} = (c\mathbb{I})_{i,\ell} = 0$ if $i \neq \ell$, we have

$$(\mathbf{E_{jk}})_{i\ell} = \langle \tilde{\pi}_{i,j}, \tilde{\pi}_{\ell,k} \rangle = 0 \text{ when } i \neq \ell.$$

And finally, if $j = k$ and $i = \ell$ we have

$$(\mathbf{E_{jj}})_{ii} = \langle \tilde{\pi}_{i,j}, \tilde{\pi}_{i,j} \rangle = c = \frac{\operatorname{tr} E_{jj}}{d} = \frac{1}{d} \text{ when } j = k \text{ and } i = \ell.$$

Summarizing:

(1) $\langle \tilde{\pi}_{i,j}, \tilde{\rho}_{\ell,k} \rangle = 0$ for all i, j, k, ℓ whenever $\rho \not\cong \pi$.

(2) $\langle \tilde{\pi}_{i,j}, \tilde{\pi}_{k,\ell} \rangle = \begin{cases} 1/d & \text{if } i = k \text{ and } j = \ell; \\ 0 & \text{otherwise.} \end{cases}$

$\square$

We now have the tools to prove our main objective.

Proposition 7.18 *The irreducible characters of a finite group G are an orthonormal basis for $\mathbb{C}_{class}[G]$, the space of class functions on G.*

Proof We will prove this proposition in two steps. First we will show that the irreducible characters are orthonormal, then pause for some useful applications, and finally show that they span the space of class functions.

Let (ρ, U) and (π, V) be irreducible finite dimensional representations of a finite group G.

Step 1: The irreducible characters are orthonormal since, for $(\rho, U) \not\cong (\pi, V)$, we have;

$$\langle \chi_U, \chi_V \rangle = \langle \tilde{\rho}_{11} + \cdots + \tilde{\rho}_{dd}, \tilde{\pi}_{11} + \cdots + \tilde{\pi}_{kk} \rangle$$
$$= \langle \tilde{\rho}_{11}, \tilde{\pi}_{11} \rangle + \langle \tilde{\rho}_{11}, \tilde{\pi}_{22} \rangle + \cdots + \langle \tilde{\rho}_{dd}, \tilde{\pi}_{kk} \rangle$$

$$= 0 + \cdots + 0$$
$$= 0,$$

and

$$\langle \chi_U, \chi_U \rangle = \langle \tilde{\rho}_{11} + \cdots + \tilde{\rho}_{dd}, \tilde{\rho}_{11} + \cdots + \tilde{\rho}_{dd} \rangle$$
$$= \langle \tilde{\rho}_{11}, \tilde{\rho}_{11} \rangle + \langle \tilde{\rho}_{11}, \tilde{\rho}_{22} \rangle + \cdots + \langle \tilde{\rho}_{dd}, \tilde{\rho}_{dd} \rangle$$
$$= 1/d + 0 + \cdots + 1/d$$
$$= 1.$$

Note that, since the trace of a matrix is basis-independent, the representations need not be unitary.

It is a standard fact from linear algebra that any orthogonal subset of a vector space is linearly independent. Therefore,

the number of inequivalent irreps = the number of irreducible characters
$$\leq \text{Dim } \mathbb{C}_{class}[G]$$
= the number of conjugacy classes.

Exercise* 7.19 Let's work through some interesting results that follow from the orthogonality relations of irreducible characters.

(1) If V is any representation of a finite group G, and if W is an irreducible representation of G, show that

$$\langle \chi_V, \chi_W \rangle = \text{ the multiplicity of } W \text{ in } V.$$

(2) If $\{V_i\}$ are irreducible representations of G, and if $V = c_1 V_1 \oplus \cdots \oplus c_k V_k$, show that

$$\langle \chi_V, \chi_V \rangle = c_1^2 + \cdots + c_k^2.$$

(3) Show that the representation V is irreducible if and only if $\langle \chi_V, \chi_V \rangle = 1$.
(4) Let χ_L be the character of the left regular representation of G on $\mathbb{C}[G]$. Show that

$$\chi_L(g) = \begin{cases} |G|, & \text{if } g = e; \\ 0, & \text{if } g \neq e. \end{cases}$$

(5) Show that every irreducible representation V_i of G appears in the left regular representation L on $\mathbb{C}[G]$, with multiplicity equal to its dimension.

(6) Show that the set $\{1/\sqrt{d_k}\,\pi_{i,j}^k\}$ spans $\mathbb{C}[G]$, and so is an orthonormal basis for $\mathbb{C}[G]$, where each π^k is an irreducible unitary representation of degree d_k.

Step 2: It remains to show that the dimension of $\mathbb{C}_{class}[G] = \mathcal{Z}\mathbb{C}[G]$ is less than or equal to the number of distinct irreducible characters of G, and so the irreducible characters span $\mathbb{C}_{class}[G]$. Toward that end, suppose that there are k mutually inequivalent irreducible representations of G, and consider the left regular representation L of G on the "first version" (Chapter 6) of the group algebra $\mathbb{C}[G]$.[1] By Maschke's theorem (Theorem 3.25) and part (5) above, $\mathbb{C}[G]$ decomposes as a direct sum of isotypic components;

$$\mathbb{C}[G] \cong r_1 W_1 \oplus \cdots \oplus r_k W_k,$$

where for each i, $r_i W_i$ is the direct sum of r_i copies of the irreducible representation W_i. What follows does not require that all of the r_i are non-zero, but by Remark 6.4 we know that they are positive.

Now if $z \in \mathcal{Z}\mathbb{C}[G]$ is central, then $zL(g)\mathbf{v} = L(g)z\mathbf{v}$ for all $g \in G$ and all $\mathbf{v} \in \mathbb{C}[G]$. That is, the map $\mathbf{v} \mapsto z.\mathbf{v}$ intertwines the left regular representation L on $\mathbb{C}[G]$ (Exercise 6.10), so by Schur's lemma there is a scalar $c_i \in \mathbb{C}$ such that $z r_i \mathbf{w}_i = c_i r_i \mathbf{w}_i$ for all $\mathbf{w}_i \in W_i$ and for each i.

Since the identity element e_G is in $\mathbb{C}[G]$, we can write

$$e_G = r_1 \mathbf{w}_1 + \cdots + r_k \mathbf{w}_k \text{ for some } \mathbf{w}_i \in W_i.$$

Thus

$$z = z e_G = z(r_1 \mathbf{w}_1 + \cdots + r_k \mathbf{w}_k) = c_1 r_1 \mathbf{w}_1 + \cdots + c_k r_k \mathbf{w}_k,$$

which shows that $\mathcal{Z}\mathbb{C}[G] = \mathbb{C}_{class}[G]$ is spanned by at most k vectors, which is the number of irreducible representations of G. Combining this with the previous result gives:

The number of inequivalent irreps of G $\leq$ Dim $\mathbb{C}_{class}[G]$
$=$ the number of conjugacy classes
$\leq$ the number of inequivalent irreps of G.

Therefore the irreducible characters span $\mathbb{C}_{class}[G]$. $\square$

---

[1] Since we have shown that both versions of the group algebra are equivalent, we can use either version as needed!

Remark 7.20 Another proof of Proposition 7.18 can be obtained by showing that any class function that is orthogonal to every irreducible character must be zero. Hence the irreducible characters are a maximal orthonormal subset of $\mathbb{C}_{class}[G]$, and so must be a basis. See [FH], Proposition 2.30.

Exercise 7.21 Illustrate the results of Propositions 7.16 and 7.18 using the matrices that you obtained in Exercise 2.34.

Exercise* 7.22 Work through the details in the above proof; that for each i there is a scalar $c_i \in \mathbb{C}$ such that $z r_i \mathbf{w}_i = c_i r_i \mathbf{w}_i$ for all $\mathbf{w}_i \in W_i$.

Exercise 7.23 Let (ρ, U) and (π, V) be representations of G, suppose $S \in \text{Hom}(U, V)$, and let $P(S) = \mathbf{S}$, where $\mathbf{S}$ is defined as in Proposition 7.16 part (1). Show that the map P is linear, and that if S is a G-map, then $P(S) = \mathbf{S} = S$. Conclude that the map $P : \text{Hom}(U, V) \to \text{Hom}_G(U, V)$ is a projection.

Remark 7.24 In the case where we have an irreducible unitary representation of $SU(2, \mathbb{C})$ or $SO(3, \mathbb{R})$, the matrix entries are called *Wigner D-functions* after the theoretical physicist Eugene P. Wigner, who was awarded the Nobel Prize in physics in 1963 for his discovery and application of symmetry principles to the study of elementary particles. The "D" stands for stands for *Darstellung*, which means "representation" in German.

7.4   Another Look

To be thorough, it is worthwhile to revisit some of the results of this chapter from a slightly different point of view. We start by establishing the dimension of the subspace $\text{Hom}_G(V, W)$.

Exercise* 7.25 Check that the action of G on $\text{Hom}(V, W)$ given by $[g.F] = gFg^{-1}$ for $F \in \text{Hom}(V, W)$, is a representation of G, and that F is a G-map if and only if F is fixed by this G-action. In other words, show that $\text{Hom}(V, W)^G = \text{Hom}_G(V, W)$ (see Exercise 2.28).

Exercise 7.26 Adapting the map from Exercise 2.28, define $\phi : \text{Hom}(V, W) \to \text{Hom}_G(V, W)$ by

$$\phi(F) := \frac{1}{|G|} \sum_{g \in G} g.F = \frac{1}{|G|} \sum_{g \in G} gFg^{-1},$$

so that ϕ is a projection. Also note that this is the map from proof of Proposition 7.16 (1). Conclude, using the results of Exercise 7.2, that

$$\text{Dim Hom}_G(V, W) = \text{tr}\,\phi = \frac{1}{|G|} \sum_{g \in G} \text{tr}(gFg^{-1}) = \frac{1}{|G|} \sum_{g \in G} \chi(g),$$

where $\chi = \chi_{\text{Hom}(V,W)}$ as in Exercise 7.25.

Our next task is to determine the character $\chi = \chi_{\text{Hom}(V,W)}$. First lets review the fact that if U is a vector space with basis $\{u_1, \ldots, u_n\}$, and if $F : U \rightarrow U$ is linear, then the diagonal entry $\tilde{F}_{k,k}$ in the matrix realization of F is the coefficient of u_k in $F(u_k)$.

Now suppose that $\{v_1, \ldots v_n\}$ is a basis for V, $\{w_1, \ldots w_m\}$ is a basis for W, and set $f_{j,i}(v_k) := \delta_{j,k} w_i$. That is, $f_{j,i}$ sends v_j to w_i, and sends all other basis vectors to $\mathbf{0}$, and hence $\{f_{j,i} \mid i = 1 \ldots m, j = 1 \ldots n\}$ is a basis for $\text{Hom}(V, W)$.

With respect to these bases, the matrix realizations are:

$$\tilde{\mathbf{v}}_i = \begin{bmatrix} 0 \\ \vdots \\ 0 \\ 1 \\ 0 \\ \vdots \\ 0 \end{bmatrix} \leftarrow i\text{th}, \quad \tilde{\mathbf{w}}_j = \begin{bmatrix} 0 \\ \vdots \\ 0 \\ 1 \\ 0 \\ \vdots \\ 0 \end{bmatrix} \leftarrow j\text{th}, \quad \text{and} \quad \tilde{f}_{j,i} = E_{i,j}.$$

We use the notation $\tilde{\mathbf{v}}_i$ and $\tilde{\mathbf{w}}_j$ rather than $\mathbf{e}_i$ and $\mathbf{e}_j$ to remind us of their source.

By the above "let's review" remark, we seek to determine the coefficient of $E_{i,j}$ in $\widetilde{g.E_{i,j}} = gf_{j,i}g^{-1}$. Since $g.f_{i,j}$ is a linear combination of the basis functions $f_{k,l}$, in terms of matrix realizations we have,

$$\tilde{g}E_{i,j}\tilde{g}^{-1} = a_{1,1}E_{1,1} + a_{1,2}E_{1,2} + \cdots + a_{i,j}E_{i,j} + \cdots + a_{m,n}E_{m,n}.$$

So we seek to determine the coefficient $a_{i,j}$. But $E_{i,j}$ sends $\tilde{\mathbf{v}}_j$ to $\tilde{\mathbf{w}}_i$, and all other basis vectors to 0, that is,

$$a_{i,j}E_{i,j}\tilde{\mathbf{v}}_j = a_{i,j}\tilde{\mathbf{w}}_i,$$

so we want the coefficient of $\tilde{\mathbf{w}}_i$ in $\tilde{g} E_{i,j} \tilde{g}^{-1} \tilde{\mathbf{v}}_j$. Using some basic linear algebra;

$$\tilde{g} E_{i,j} \tilde{g}^{-1} \tilde{\mathbf{v}}_j = \tilde{g} E_{i,j} \left(\sum_k \tilde{g}_{j,k}^{-1} \tilde{\mathbf{v}}_k \right)$$

$$= \tilde{g} \sum_k \tilde{g}_{j,k}^{-1} E_{i,j} (\tilde{\mathbf{v}}_k)$$

$$= \tilde{g} \, \tilde{g}_{j,j}^{-1} \, \tilde{\mathbf{w}}_i$$

$$= \tilde{g}_{j,j}^{-1} \, g . \tilde{\mathbf{w}}_i$$

$$= \tilde{g}_{j,j}^{-1} \sum_k \tilde{g}_{i,k} \, \tilde{\mathbf{w}}_k .$$

Therefor, the coefficient of $\tilde{\mathbf{w}}_i$ is $\tilde{g}_{i,i} \, \tilde{g}_{j,j}^{-1}$. It follows that

$$\chi(g) = \sum_{i,j} a_{i,j} = \sum_{i,j} \tilde{g}_{i,i} \tilde{g}_{j,j}^{-1} = \sum_i \tilde{g}_{i,i} \sum_j \tilde{g}_{j,j}^{-1} = \chi_V(g) \chi_W(g^{-1}).$$

Exercise* 7.27 Show that $\chi_W(g^{-1}) = \overline{\chi_W(g)}$. Hint: W is a complex vector space (so we are guaranteed non-zero eigenvectors), and $|G|$ is finite.

To summarize,

$$\chi(g) = \chi_{\mathrm{Hom}(V,W)}(g) = \chi_V(g) \overline{\chi_W(g)}.$$

It is therefore natural to define an inner product on characters by

$$\langle \chi_V, \chi_W \rangle := \frac{1}{|G|} \sum_{g \in G} \chi_V(g) \overline{\chi_W(g)}$$

as in Example 3.32, and the whole discussion yields

$$\langle \chi_V, \chi_W \rangle := \frac{1}{|G|} \sum_{g \in G} \chi_V(g) \overline{\chi_W(g)} = \mathrm{tr}\, \phi = \mathrm{Dim}\, \mathrm{Hom}_G(V, W).$$

It follows that if V is irreducible, then $\langle \chi_V, \chi_W \rangle$ is the multiplicity of V in W.

7.5 Hints and Additional Comments

Exercise 7.2 Let $\{v_1, \ldots, v_k\}$ a basis for U which we can extend to $\{v_1, \ldots, v_k, \ldots, v_n\}$ to obtain a basis for V. With respect to this basis, the matrix realization for P_U consists of ones for the first k diagonal entries and zeroes elsewhere. Alternatively, each basis vector for U is an eigenvector for P_U with eigenvalue one, all other basis vectors for V have eigenvalue zero.

Furthermore, if U is a G-invariant subspace of V, then by Maschke's theorem there is a complementary subspace W such that $V = U \oplus W$, and P_U acts as the identity on U and is the zero-map on W.

Exercise 7.5 Observe that if $f : G \to \mathbb{F}$ is a class function, then $f(h) = \sum f(h)\delta_{[g]}(h)$, the sum being taken over each conjugacy class $[g]$ in G.

Exercise 7.12 The action of S_3 permutes the distinct monomials that are a basis for $\mathcal{P}_{(2,1,0)}$. The character of S_3 acting on $\mathcal{P}_{(2,1,0)}$ is the number of basis monomials fixed by each element of S_3. The identity fixes all six basis monomials, the other group elements fix none.

Exercise 7.13 Take one representative from each conjugacy class, then solve the 3×3 system:

$$r_1 \chi_\mathcal{I}(1) + r_2 \chi_\mathcal{W}(1) + r_3 \chi_\mathcal{A}(1) = 0$$
$$r_1 \chi_\mathcal{I}(1,2) + r_2 \chi_\mathcal{W}(1,2) + r_3 \chi_\mathcal{A}(1,2) = 0$$
$$r_1 \chi_\mathcal{I}(1,2,3) + r_2 \chi_\mathcal{W}(1,2,3) + r_3 \chi_\mathcal{A}(1,2,3) = 0$$

to obtain $r_1 = r_2 = r_3 = 0$.

Exercise 7.19 (1) If V contains an irreducible representation equivalent to W, then

$$V \cong c_1 W \oplus c_2 V_2 \oplus \cdots \oplus c_k V_k \quad \text{and} \quad \chi_V = c_1 \chi_W + c_2 \chi_{V_2} + \cdots + c_k \chi_{V_k},$$

where each of the V_i is inequivalent to W. Therefore, by orthonormality of characters;

$$
\begin{aligned}
\langle \chi_V, \chi_W \rangle &= \langle c_1 \chi_W + c_2 \chi_{V_2} + \cdots + c_k \chi_{V_k}, \chi_W \rangle \\
&= c_1 \langle \chi_W, \chi_W \rangle + c_2 \langle \chi_{V_2}, \chi_W \rangle + \cdots + \langle \chi_{V_k}, \chi_W \rangle \\
&= c_1 \langle \chi_W, \chi_W \rangle + 0 + \cdots + 0 \\
&= c_1 \\
&= \text{the multiplicity of } W \text{ in } V.
\end{aligned}
$$

Item (2) is proven similarly.

To show (3), by part (2) we have,

$$\langle \chi_V, \chi_V \rangle = c_1^2 + \cdots + c_k^2 = 1,$$

so we must have some c_i equal to 1 and all other constants equal to 0.

(4) Choose the set $\{\mathbf{g}\}_{g \in G}$ as a basis for $\mathbb{C}[G]$. Then, since $gh \neq h$ unless $g = e$, the diagonal entries of the matrix realization of a group element $g \in G$ are all equal to one for $g = e$, and all equal to zero otherwise (Exercise 7.6).

(5)

The multiplicity of V_i in $L = \langle \chi_L, \chi_{V_i} \rangle$

$$
\begin{aligned}
&= \tfrac{1}{|G|} \sum_{g \in G} \chi_L(g) \chi_{V_i}(g^{-1}) \\
&= \tfrac{1}{|G|} \chi_L(e) \chi_{V_i}(e) && \text{(since } \chi_L(g) = 0 \text{ for } g \neq e \text{)} \\
&= \tfrac{1}{|G|} |G| \chi_{V_i}(e) && \text{(since } \chi_L(e) = |G| \text{)} \\
&= \text{Dim } V_i && \text{(since } \chi_{V_i}(e) = \text{Dim } V_i \text{)}.
\end{aligned}
$$

As a further result,

$$|G| = \text{Dim } \mathbb{C}[G] = \sum (\text{Dim } V_i)^2,$$

where the sum is taken over all irreducible representations V_i of G.

(6) Since for each irreducible unitary representation V_k there are $(\text{Dim } V_k)^2$ orthogonal matrix entries, part (5) implies that the set $\{1/\sqrt{d_k}\, \tilde{\pi}_{i,j}^k\}$ spans $\mathbb{C}[G]$, and so is an orthonormal basis for $\mathbb{C}[G]$, where each π^k is an irreducible unitary representation of degree d_k.

Exercise 7.22 We have that z is central in $\mathbb{C}[G]$, and

$$\mathbb{C}[G] \cong r_1 W_1 \oplus \cdots \oplus r_k W_k,$$

where for each i, $r_i W_i$ is the W_i isotypic component of $\mathbb{C}[G]$. Since each copy W_i' of W_i is G-invariant, $z.\mathbf{w}_i \in W_i'$ for all $\mathbf{w}_i \in W_i'$, and so $z.r_i W_i \cong r_i z.W_i$. Since each W_i is irreducible, $z.\mathbf{w}_i = c_i \mathbf{w}_i$ for some scalar c_i by the strong version of Schur's lemma.

A somewhat subtle point is that the SAME constant c_i works for each equivalent copy of W_i. That is, if $W_i \cong W_i'$, so that $z.\mathbf{w}_i = c_i \mathbf{w}_i$, and $z.\mathbf{w}_i' = c_i' \mathbf{w}_i'$, then $c_i = c_i'$. Now since $W_i \cong W_i'$, there is a G-isomorphism $\phi \colon W_i \to W_i'$. Let $\mathbf{w}_i' = \phi(\mathbf{w}_i)$. Then

$$c_i' \mathbf{w}_i' = z.\mathbf{w}_i' = z.\phi(\mathbf{w}_i) = \phi(z.\mathbf{w}_i) = \phi(c_i \mathbf{w}_i) = c_i \phi(\mathbf{w}_i) = c_i \mathbf{w}_i'.$$

Thus $c_i = c_i'$.

Exercise 7.25 Let $F \in \text{Hom}(V, W)$, let $\mathbf{v} \in V$, and let $g, h \in G$. Then the definition says that

$$[g.F](\mathbf{v}) := g.[F(g^{-1}.\mathbf{v})].$$

It's worth emphasising that the notation $[g.F]$ on the left side of the ":=" means the action of G on $\text{Hom}(V, W)$. On the right side of the ":=," the notation $g^{-1}.\mathbf{v}$ means the action of G on V, while the notation $g.[\quad]$ means the action of G on W. You might try rewriting this specifying the representations using ρ, π, etc.

To verify that this is a representation, first check that if $F \in \text{Hom}(V, W)$, then $g.F \in \text{Hom}(V, W)$ for all $g \in G$. Now let $g, h \in G$, and let $\mathbf{v} \in V$. Then

$$[h.[g.F]](\mathbf{v}) = h.\Big[[g.F](h^{-1}\mathbf{v})\Big] = h.[g.F(g^{-1}h^{-1}.\mathbf{v})] = hg.[F((gh)^{-1}.\mathbf{v}) = [hg.F](\mathbf{v}).$$

Now if F is a G-map, then

$$gF(g^{-1}\mathbf{v}) = F(gg^{-1}(\mathbf{v}) = F(\mathbf{v}) \text{ for all } \mathbf{v} \in V,$$

so $F \in \text{Hom}^G(V, W)$.

Conversely, if $F \in \text{Hom}^G(V, W)$, then

$$gF(g^{-1}\mathbf{v}) = F(\mathbf{v}) \text{ for all } \mathbf{v} \in V,$$

so $F(g^{-1}\mathbf{v}) = g^{-1}F(\mathbf{v})$, and F is a G-map.

Exercise 7.27 Since W is a complex vector space, the action of $g \in G$ has an eigenvector $\mathbf{w}$ with eigenvalue $c \in \mathbb{C}$. That is, $g.\mathbf{w} = c\mathbf{w}$. Since G is finite, $g^k = e$ for some positive integer k (Exercise 1.50), and therefore $e.\mathbf{w} = g^k.\mathbf{w} = c^k\mathbf{w} = \mathbf{w}$, so $c^k = 1$, $|c|^2 = c\bar{c} = 1$, and hence $c^{-1} = \bar{c}$. Since χ is the sum of the eigenvalues, the result follows.

The Irreducible Representations of S_n: Young Symmetrizers

8

In this chapter we explicitly construct the irreducible representations of S_n in the group algebra $\mathbb{C}[S_n]$. For each partition of n we define an element in $\mathbb{C}[S_n]$, called a Young symmetrizer, that generates an irreducible representation of S_n in $\mathbb{C}[S_n]$. Consequently, there is a one-to-one correspondence between the irreducible representations of S_n and the integer partitions of n. Furthermore, given any representation V of S_n, we can use these symmetrizers as projections to obtain irreducible representations in V. Part of the discussion here follows [FH], Lecture 4 and [E], Section 5.12.

Young symmetrizers, Young diagrams, etc. are named after Alfred Young, a British mathematician who wrote a long series of papers on the symmetric group while serving as a clergyman. A modern presentation of much of his work is given in [R].

8.1 Partitions Again: Young Tableaux

We have seen, via character theory, how irreducible representations of S_n correspond to the conjugacy classes of S_n, which in turn correspond to partitions of n. For each $\lambda = (\lambda_1, \lambda_2, \ldots, \lambda_k) \vdash n$ we associate the *Young diagram of shape* λ, which is k rows of boxes, left justified, with λ_1 boxes in the first row, λ_2 boxes in the second row, etc. We will use a label such as λ interchangeably to denote both the partition and the diagram. For example,

$$\lambda = (4, 2, 2, 1) = $$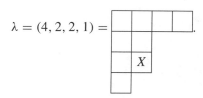

Boxes in a Young diagram are labeled like entries in a matrix; the above diagram has an X in the 3, 2 position. Young diagrams are sometimes referred to as *Ferrers diagrams*, and are sometimes depicted "upside down," especially by Francophones.

For any partition λ we can obtain the *conjugate partition* λ' by interchanging the rows and columns of λ. For example, if $\lambda = (4, 2, 2, 1)$, then

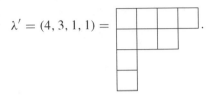

A *tableau* is a Young diagram whose boxes are filled with entries taken from an *alphabet* (any totally ordered set). A *Young tableau T* is a Young diagram whose n boxes are filled with the numbers $1, \ldots, n$ without repetition. We will use the notation T^λ if we need to specify that T has shape λ. A *standard Young tableau* is a Young tableau where the entries in the rows are increasing left-to-right, and where the entries in the columns are increasing top-to-bottom. For example, the tableau

$$
\begin{array}{|c|c|c|c|}
\hline
1 & 3 & 5 & 7 \\
\hline
2 & 4 \\
\cline{1-2}
6 & 8 \\
\cline{1-2}
9 \\
\cline{1-1}
\end{array}
$$

is standard.

The symmetric group acts on a Young tableau in the obvious way, by permuting the entries in the tableau. For example,

$$
(1, 3, 4). \begin{array}{|c|c|c|c|}
\hline
1 & 3 & 5 & 7 \\
\hline
2 & 4 \\
\cline{1-2}
6 & 8 \\
\cline{1-2}
9 \\
\cline{1-1}
\end{array}
=
\begin{array}{|c|c|c|c|}
\hline
3 & 4 & 5 & 7 \\
\hline
2 & 1 \\
\cline{1-2}
6 & 8 \\
\cline{1-2}
9 \\
\cline{1-1}
\end{array}.
$$

Exercise 8.1 Show that S_n acts transitively on the set of all Young tableaux with the same shape. Consequently, if two tableaux T and T' have the same shape, then $T' = \sigma.T$ for some $\sigma \in S_n$.

8.2   Orderings on Partitions

There are different ways to order partitions of a fixed integer n. One common and useful way is the lexicographic, or "dictionary" ordering.

Definition 8.2 Let $\lambda = (\lambda_1, \ldots, \lambda_n)$ and $\mu = (\mu_1, \ldots, \mu_n)$ be two integer partitions of n. Then $\lambda > \mu$ in the *lexicographic order* if the first non-zero $\lambda_i - \mu_i$ is positive. Note that this is a *total order* on partitions; any partitions $\lambda, \mu \vdash n$ obey exactly one of $\lambda > \mu$, $\lambda = \mu$, or $\mu > \lambda$. For example, with $n = 4$ we have

$$(4) > (3, 1) > (2, 2) > (2, 1, 1) > (1, 1, 1, 1).$$

To be thorough, we also define the dominance order.

Definition 8.3 Let $\lambda = (\lambda_1, \ldots, \lambda_n)$ and $\mu = (\mu_1, \ldots, \mu_n)$ be two integer partitions of n. Then $\lambda \trianglerighteq \mu$ in the *dominance order* if

$$\lambda_1 \geq \mu_1, \quad \lambda_1 + \lambda_2 \geq \mu_1 + \mu_2, \quad \ldots, \quad \lambda_1 + \lambda_2 + \ldots + \lambda_n \geq \mu_1 + \mu_2 + \ldots + \mu_n.$$

This is only a partial order on partitions. For example, for $n = 6$ we have $(5, 1) \trianglerighteq (4, 2)$, but $(4, 1, 1)$ and $(3, 3)$ are not comparable. However, $(4, 1, 1) \trianglerighteq (3, 2, 1)$, and $(3, 3) \trianglerighteq (3, 2, 1)$.

8.3 Young Symmetrizers

Given a Young tableau T associated to a partition $\lambda \vdash n$, define two subgroups of S_n as follows: the *horizontal subgroup* (sometimes called the *row stabilizer*),

$$H_T := \{h \in S_n \mid h \text{ setwise preserves each row of } T\};$$

and the *vertical subgroup* (sometimes called the *column stabilizer*)[1],

$$V_T := \{v \in S_n \mid v \text{ setwise preserves each column of } T\}.$$

For example, if

$$T = \begin{array}{|c|c|c|} \hline 1 & 2 & 4 \\ \hline 3 & 5 \\ \cline{1-2} \end{array},$$

then $(1, 2, 4)(3, 5) \in H_T$ and $(1, 3)(2, 5) \in V_T$. Elements in H_T and V_T are called, respectively, *row permutations* and *column permutations*.

---

[1] The terms vertical or horizontal stabilizer would also make sense, but sound like parts of an aircraft.

We now construct elements in the group algebra $\mathbb{C}[S_n]$. For the subgroup H_T we form the *row symmetrizer*,

$$\mathcal{H}_T := \sum_{h \in H_T} h,$$

and for the subgroup V_T we form the *column anti-symmetrizer*,

$$\mathcal{V}_T := \sum_{v \in V_T} \mathrm{sgn}(v)v.$$

We then construct the *Young symmetrizer* $\mathbf{E}_T$ (sometimes called a *Young projector*) for the given tableau T,

$$\mathbf{E}_T := \mathcal{H}_T \mathcal{V}_T = \sum_{h \in H_T} h \sum_{v \in V_T} \mathrm{sgn}(v)v = \sum_{h \in H_T} \sum_{v \in V_T} h\, \mathrm{sgn}(v)v,$$

which is likewise an element in the group algebra. We can think of $\mathbf{E}_T$ as symmetrizing along the rows of T and anti-symmetrizing down the columns.

Example 8.4 For the tableau

$$T = \begin{array}{|c|c|} \hline 1 & 2 \\ \hline 3 \\ \cline{1-1} \end{array},$$

we have $H_T = \{(1), (1, 2)\}$, $V_T = \{(1), (1, 3)\}$, and

$$\mathbf{E}_T = [(1) + (1, 2)] \times [(1) - (1, 3)] = (1) + (1, 2) - (1, 3) - (1, 3, 2).$$

Notation 8.5 In some cases it will be more applicable to use the notation $H_\lambda, \mathcal{V}_\lambda, \mathbf{E}_\lambda$, etc., when it is important to emphasize the partition λ rather than the numbering of T^λ.

Remark 8.6 Some authors [JK, Section 3.1] write Young symmetrizers "backwards," as $\mathbf{E}_T^* := \mathcal{V}_T \mathcal{H}_T$. It should be apparent that for this other version of a Young symmetrizer the following proofs carry through with the necessary adjustments. The definition we present here seems to be more common in the literature. See also Exercise 8.39.

In these next few exercises we develop some properties of H_T, V_T, $\mathcal{H}_T$, $\mathcal{V}_T$, and $\mathbf{E}_T$ for later application.

Exercise 8.7 Write out some examples of standard Young tableaux T for different integer partitions of various n, and verify that H_T and V_T are subgroups of S_n.

Exercise 8.8 Verify that $H_T \cap V_T = \{(1)\}$, and thus every element σ in the set

$$H_T V_T := \{hv \mid h \in H_T, v \in V_T\}$$

has a unique representation as $\sigma = hv$. Hint: See Exercise 2.15. Also check that $H_T V_T$ need not be a subgroup of S_n.

Exercise 8.9 For any Young tableau T, verify that the identity element of S_n appears in $\mathbf{E}_T$ with coefficient one.

Exercise* 8.10 Let T and T' be Young tableau with the same shape but with different numberings. Show that H_T and $H_{T'}$ are conjugate as subgroups of S_n, and similarly for the vertical subgroups. Hint: $T' = \sigma.T$ for some $\sigma \in S_n$, so we need to show that

- $H_{\sigma.T} = \sigma H_T \sigma^{-1}$, and that
- $V_{\sigma.T} = \sigma V_T \sigma^{-1}$.

Conclude that $\mathbf{E}_{\sigma.T} = \sigma \mathbf{E}_T \sigma^{-1}$.

Exercise* 8.11 Generalize Exercise 8.10; show that if a group G acts on a set S, and if s and s' are in the same G-orbit, then the stabilizer subgroups (Exercise 1.62) of s and s' are conjugate. That is, $G_{s'} = xG_s x^{-1}$ for some $x \in G$.

Exercise 8.12 Check that if $h \in H_T$, and if $v \in V_T$, then

- $h\mathcal{H}_T = \mathcal{H}_T h = \mathcal{H}_T$, and
- $v\mathcal{V}_T = \mathcal{V}_T v = \mathrm{sgn}(v)\mathcal{V}_T$.

Hint: Use the fact that summing over all $h \in H$ is equivalent to summing over $h'h$, for a fixed $h' \in H$ and all $h \in H$.

Definition 8.13 An element $\mathfrak{e}$ in an algebra A such at $\mathfrak{e}^2 = \mathfrak{e}$ is said to be *idempotent* in A. An element $\mathfrak{e}$ such that $\mathfrak{e}^2 = \kappa\mathfrak{e}$ for some constant κ is said to be *essentially idempotent* in A. For example, the numbers 1 and 0 are obviously idempotents in $\mathbb{R}$. The matrix $\left(\begin{smallmatrix} 2 & -1 \\ 2 & -1 \end{smallmatrix}\right)$ is an idempotent in the algebra of 2×2 matrices. A useful feature of idempotents is that they can be used as projections.

Exercise* 8.14 Show that $\mathcal{V}_T^2 = |V_T|\mathcal{V}_T$ and that $\mathcal{H}_T^2 = |H_T|\mathcal{H}_T$. Conclude that $\mathcal{V}_T$ and $\mathcal{H}_T$ are relative idempotents, and that $\frac{1}{|V_T|}\mathcal{V}_T$ and $\frac{1}{|H_T|}\mathcal{H}_T$ are projections. Describe the spaces onto which they project.

Exercise* 8.15 By Remark 6.3, if V is a representation of S_n, then the Young symmetrizers, as elements of $\mathbb{C}[S_n]$, also act on V. For the tableau T given in Example 8.4:

(1) Show that $\mathbf{E}_T^2 = \kappa \mathbf{E}_T$ for some constant κ. Conclude that if $\widehat{\mathbf{E}}_T := \frac{1}{\kappa} \mathbf{E}_T$, then $\widehat{\mathbf{E}}_T^2 = \widehat{\mathbf{E}}_T$, and hence $\widehat{\mathbf{E}}_T : V \to V$ is a projection.
(2) Consider $\widehat{\mathbf{E}}_T : \mathcal{P}_1 \to \mathcal{P}_1$. Let $\widehat{\mathbf{E}}_T$ act on the basis monomials $\{\mathbf{x}, \mathbf{y}, \mathbf{z}\}$. Conclude that the image of $\widehat{\mathbf{E}}_T$ in $\mathcal{P}_1$ lies in the irreducible subspace $\mathcal{W}$ from Example 2.29.

Exercise 8.16 Write out some standard Young tableaux for partitions of 3 and 4, and show that $\mathbf{E}_T^2 = \kappa \mathbf{E}_T$ for some constant κ.

The symmetrizers $\widehat{\mathbf{E}}_T$ and $\mathbf{E}_T$ are, respectively, idempotent and essentially idempotent, and can be used as projections, as in Sect. 2.5. The following exercises work through some examples.

Exercise* 8.17 Compute the Young symmetrizers for the standard Young tableaux corresponding to the two partitions $\lambda \vdash 2$. Have them act on the basis monomials in $\mathcal{P}_1(x, y)$. Interpret the results.

Exercise* 8.18 Compute the Young symmetrizer for the standard Young tableau $\boxed{1\;2\;3}$, and have it act on the basis monomials in $\mathcal{P}_1(x, y, z)$. Interpret the results.

Exercise* 8.19 Compute the Young symmetrizer for the standard Young tableau

$$\boxed{\begin{array}{c} 1 \\ 2 \\ 3 \end{array}},$$

and have it act on the basis monomials in $\mathcal{P}_1(x, y, z)$. Interpret the results. Also have it act on $\mathcal{P}_{(2,1,0)}(x, y, z) \subseteq \mathcal{P}_3(x, y, z)$ and interpret the results.

Exercise* 8.20 More projections using symmetrizers.

(1) Let $\widehat{\mathbf{E}}_{T_1}$ be the Young symmetrizer for the standard Young tableau

$$T_1 = \boxed{\begin{array}{cc} 1 & 2 \\ 3 & \end{array}}.$$

Set $\mathbf{v}_1 = \widehat{\mathbf{E}}_{T_1}.(x^2 y)$, and set $\mathbf{v}_2 = (1, 3).\mathbf{v}_1$. Show that $\{\mathbf{v}_1, \mathbf{v}_2\}$ spans an irreducible representation of S_3 in $\mathcal{P}_{(2.1.0)}$.

(2) Let $\widehat{\mathbf{E}}_{T_2}$ be the Young symmetrizer for the standard Young tableau

$$T_2 = \begin{array}{|c|c|} \hline 1 & 3 \\ \hline 2 \\ \cline{1-1} \end{array} .$$

Set $\mathbf{w}_1 = \widehat{\mathbf{E}}_{T_2}.(x^2 y)$, and set $\mathbf{w}_2 = (1, 2).\mathbf{w}_1$. Show that $\{\mathbf{w}_1, \mathbf{w}_2\}$ spans an irreducible representation of S_3 in $\mathcal{P}_{(2.1.0)}$ that is independent from the irreducible representation spanned by$\{\mathbf{v}_1, \mathbf{v}_2\}$. That is, show that the set $\{\mathbf{v}_1, \mathbf{v}_2, \mathbf{w}_1, \mathbf{w}_2\}$ is linearly independent.

(3) Let $\widehat{\mathbf{E}}_{T_1}$ and $\widehat{\mathbf{E}}_{T_2}$ act on $\mathcal{P}_1$ as in Exercise 8.15. Interpret the results.

The above exercises illustrate the general case. Given a representation V of S_n, for every $\lambda \vdash n$, and for each standard Young tableau T with shape λ, we obtain a Young symmetrizer $\widehat{\mathbf{E}}_T : V \to V$ which is a projection. Given any vector $\mathbf{v} \in V$, the vector $\mathbf{v}_T := \widehat{\mathbf{E}}_T.\mathbf{v}$ *generates* an irreducible representation of S_n. That is, the set $\{\sigma.\mathbf{v}_T \mid \sigma \in S_n\}$ spans an irreducible representation of S_n. If the space V is "large enough," the Young symmetrizers corresponding to different standard Young tableau for the same partition λ will project onto different copies of the same irreducible representation, as in Exercise 8.20. Section 8.5 provides more details.

8.4   Construction of Irreducible Representations in $\mathbb{C}[S_n]$

We have seen some simple examples of irreducible representations of the symmetric group in the familiar landscape of polynomial spaces. A natural vector space on which S_n acts is its own group algebra, $\mathbb{C}[S_n]$, in which we will construct all of the irreducible representations of S_n.

Definition 8.21 For a Young tableau $T^\lambda = T$, define

$$V^\lambda := \mathbb{C}[S_n]\mathbf{E}_T = \{a\mathbf{E}_T \mid a \in \mathbb{C}[S_n]\}.$$

Since any $\sigma \in S_n$ is also in $\mathbb{C}[S_n]$, it should be clear that V^λ is an S_n-invariant subspace of $\mathbb{C}[S_n]$ under left multiplication. Some authors call V^λ the *Specht module* associated to the partition λ, but the classical construction of Specht modules is on the space of "polytabloids" given in Chap. 11.

The primary result of this chapter is articulated in the following theorem.

Theorem 8.22 *Let $\lambda \vdash n$, and let T be a Young tableau with shape λ. Then $V^\lambda :=$ $\mathbb{C}[S_n]\mathbf{E}_T$ is an irreducible representation of S_n under left multiplication. If $\lambda \neq$*

μ, then V^λ and V^μ are inequivalent representations of S_n, and every irreducible representation of S_n is equivalent to V^λ for some λ.

Remark* 8.23 The *Frobenius character formula* provides a way of computing the character of an irreducible representation of S_n with a given signature λ. See [Si], Section VI.5, or [FH], Formula 4.10. The formula realizes characters as coefficients of a generating function.

Our goal will be to show that if λ and μ are partitions of n, and if V^λ and V^μ are as in Definition 8.21, then the dimension of $\mathrm{Hom}_{\mathbb{C}[S_n]}(V^\lambda, V^\mu)$ is equal to 0 or 1. In the first case, the representations are inequivalent by Corollary 3.17. In the second case, the representations are equivalent and irreducible by Corollary 3.18.

The proof requires that we first develop more tools. We start with an exercise that illustrates the proposition that follows.

Exercise* 8.24 For the tableau T in Example 8.4, compute $\mathcal{H}_T \sigma \mathcal{V}_T$ for the following values of $\sigma \in S_3$.

(1) $\sigma = (1, 3, 2)$. Note that $(1, 3, 2) = (1, 2)(1, 3) \in H_T V_T$.
(2) $\sigma = (2, 3)$. Note that $(2, 3) \notin H_T V_T$.
(3) $\sigma = (1, 2, 3)$. Note that $(1, 2, 3) = (1, 3)(1, 2) \notin H_T V_T$.
(4) Try some similar computations with tableaux for partitions of 4.

Proposition 8.25 *Let T be a standard Young tableau. If $a \in \mathbb{C}[S_n]$, then*

$$\mathcal{H}_T a \mathcal{V}_T = \ell(a) \mathcal{H}_T \mathcal{V}_T = \ell(a) \mathbf{E}_T, \quad \text{where} \quad \ell \colon \mathbb{C}[S_n] \to \mathbb{C} \text{ is linear.}$$

Proof It is sufficient to show this for any $a = \sigma \in S_n$, and then extend by linearity. There are two cases to consider:

(1) If $\sigma \in H_T V_T$, then by Exercise 8.8 σ has a unique representation as $\sigma = hv$ for $h \in H_T$ and $v \in V_T$. Consequently,

$$\mathcal{H}_T \sigma \mathcal{V}_T = \mathcal{H}_T hv \mathcal{V}_T = \mathrm{sgn}(v) \mathcal{H}_T \mathcal{V}_T = \pm \mathbf{E}_T \quad \text{(by Exercise 8.12)}.$$

(2) Let T be a standard Young tableau, let $\sigma \in S_n$, and set $T' = \sigma T$. If T and T' have a pair $i, j \in \{1, \ldots, n\}$ with both i and j in the same row of T, and with both i and j in the same column of T', then the transposition $t = (i, j) \in H_T \cap V_{T'}$. But, by Exercise 8.10, $t \in V_{T'}$ is equivalent to saying that $\sigma^{-1} t \sigma \in$

V_T. Therefore,

$$
\begin{aligned}
\mathcal{H}_T \sigma V_T &= \mathcal{H}_T t \sigma V_T && \text{(since } t \in H_T \text{ and Exercise 8.12)} \\
&= \mathcal{H}_T \sigma (\sigma^{-1} t \sigma) V_T && \text{(insert } \sigma \sigma^{-1} \text{ and regroup)} \\
&= -\mathcal{H}_T \sigma V_T && \text{(since } \sigma^{-1} t \sigma \in V_T \text{ and} \\
& && \text{sgn}(\sigma^{-1} t \sigma) = \text{sgn}(t) = -1).
\end{aligned}
$$

Hence $\mathcal{H}_T \sigma V_T$ must be equal to zero.

(3) There is no third case; we claim that if there is no such pair i, j as in case (2), then $\sigma \in H_T V_T$ and we are back to case (1).

It is useful here to work through a simple exercise.

Exercise* 8.26 Consider the two tableaux

$$
T = \begin{array}{|c|c|c|}
\hline
1 & 2 & 3 \\
\hline
4 & 5 & 6 \\
\hline
7 \\
\cline{1-1}
\end{array}
\quad \text{and} \quad
T' = \begin{array}{|c|c|c|}
\hline
3 & 4 & 1 \\
\hline
7 & 2 & 6 \\
\hline
5 \\
\cline{1-1}
\end{array}
= (4,5)(1,3)(2,5)(4,7).T,
$$

and observe that there is no pair i, $j \in \{1, \ldots, 7\}$ with both i and j in the same row of T, and also with both i and j in the same column of T'.

(1) Find a permutation $v_1' \in V_{T'}$ so that T and $v_1'.T'$ have the same entries in the first row (but possibly in a different order).

(2) Now find a permutation $h_1 \in H_T$ so that $h_1.T$ and $v_1'.T'$ have identical first rows.

(3) Confirm that the tableaux T and $h_1.T$ have the same horizontal subgroups, and that the tableaux T' and $v_1'.T'$ have the same vertical subgroups. Conclude that we still have the condition that there is no pair i, j with both i and j in the same row of $h_1.T$, and with both i and j also in the same column of $v_1'.T'$.

(4) Repeat the above steps on the second row of $h_1.T$ and $v_1'.T'$. That is, find $h_2 \in H_T$ and $v_2' \in V_{T'}$ so that $h_2 h_1.T$ and $v_2' v_1'.T'$ have identical first and second rows. Observe that there is no need to consider the third rows.

(5) We now have

$$
h_2 h_1.T = v_2' v_1'.T', \text{ or equivalently } \sigma.T = (v_2' v_1')^{-1} h_2 h_1.T = T'.
$$

Finally, check that $\sigma = (v_2' v_1')^{-1} h_2 h_1 \in H_T V_T$.

Continuing with our proof, we generalize the above example; given any standard Young tableau T and any $\sigma \in S_n$, set $T' = \sigma T$. Then—provided that T and T' DO NOT have a pair i, $j \in \{1, \ldots, n\}$ with both i and j in the same row of T and both i and j in the same column of T'—we can find permutations $h \in H_T$ and $v' \in V_{T'}$ so that $hT = v'T'$, or equivalently $(v')^{-1} hT = T' = \sigma.T$.

We now show that $\sigma = (v')^{-1}h \in H_T V_T$. Multiplying by the "identity in disguise" hh^{-1} on the left and regrouping yields

$$(v')^{-1}h = (hh^{-1})(v')^{-1}h = h[h^{-1}(v')^{-1}h].$$

Thus it remains to establish that $[h^{-1}(v')^{-1}h] \in V_T$.

Note first that, since $T = \sigma^{-1}.T'$, we have $V_T = V_{\sigma^{-1}.T'} = \sigma^{-1}V_{T'}\sigma$ by Exercise 8.10. Therefore

$$\begin{aligned}
h^{-1}(v')^{-1}h &= h^{-1}[v'(v')^{-1}](v')^{-1}h && \text{(insert } v'(v')^{-1}) \\
&= [h^{-1}v'](v')^{-1}[(v')^{-1}h] && \text{(regroup)} \\
&= \sigma^{-1}(v')^{-1}\sigma && \text{(since } \sigma = (v')^{-1}h) \\
&\in V_T. && \text{(since } (v')^{-1} \in V_{T'})
\end{aligned}$$

$\square$

For future reference, we record two related results.

Proposition 8.27 *Let T and T' be Young tableaux with the same shape. Then either there are two distinct positive integers that occur in the same row of T and in the same column of T', or there is some $h \in H_T$ and some $v' \in V_{T'}$ such that $hT = v'T'$.*

Proof The proof is contained in the proof of Proposition 8.25. $\square$

This next exercise illustrates the proposition that follows.

Exercise 8.28 Choose two Young tableaux T and T' with shapes λ and μ respectively, and with $\lambda > \mu$ in the lexicographic order. Find a pair of integers i, j that appear in the same row of T, and that also appear in the same column of T'.

Proposition 8.29 *Let $\lambda = (\lambda_1, \lambda_2, \ldots, \lambda_n)$ and $\mu = (\mu_1, \mu_2, \ldots, \mu_n)$ be partitions of n (possibly with trailing zeros), and suppose that T^λ and T^μ are standard Young tableau with shapes λ and μ respectively. Let $\mathcal{H}_{T^\lambda}$ be the row symmetrizer for T^λ, and let $\mathcal{V}_{T^\mu}$ be the column anti-symmetrizer for T^μ. If $\lambda > \mu$ in the lexicographic order, then $\mathcal{H}_{T^\lambda}\mathbb{C}[S_n]\mathcal{V}_{T^\mu} = 0$.*

Proof Again, by linearity it is sufficient to show that for each $\sigma \in S_n$ we have $\mathcal{H}_{T^\lambda}\sigma\mathcal{V}_{T^\mu} = 0$. Set $T = T^\lambda$, and set $T' = \sigma T^\mu$. We will show that there MUST be a pair of integers i, j that appear in the same row of T and that also appear in the same column of T'. Consequently, there is a transposition $t = (i, j) \in H_{T^\lambda}$ such that $\sigma^{-1}t\sigma \in V_{T^\mu}$, and the result follows as in the proof of Proposition 8.25 (2).

If $\lambda_1 > \mu_1$, then the distinct entries in the first row of T can't lie in distinct columns of T' since they must be distributed among fewer available columns. Thus two of the entries from the first row of T must lie in the same column of T' (by the pigeonhole principle) and we are done.

If $\lambda_1 = \mu_1$, then either there is a pair of entries in the same row of T and the same column of T', in which case $\mathcal{H}_{T^\lambda}\sigma V_{T^\mu} = 0$, or we can find a row permutation $h_1 \in H_T$ and a column permutation $v'_1 \in V_{T'}$ such that $h_1 T$ and $v'_1 T'$ have the same first row. So now we consider the second row of $h_1 T$ and $v'_1 T'$, and play the same game.

We continue in this manner until we reach the case where $\lambda_k > \mu_k$ for some k, which is guaranteed because $\lambda > \mu$ in the lexicographic order, and in which case there must be a pair of integers i, j that appear in the same row of T and that also appear in the same column of T'. □

The next proposition generalizes Exercises 8.15 and 8.16.

Proposition 8.30 *Let T be a Young tableau with shape λ. Let $\mathbf{E}_T$ be the Young symmetrizer corresponding to T, let V^λ be as in Definition 8.21, and let $d = \dim V^\lambda$. Then $\mathbf{E}_T^2 = \kappa \mathbf{E}_T$, where $\kappa = \frac{n!}{d}$. It follows that $\widehat{\mathbf{E}}_T = \frac{1}{\kappa}\mathbf{E}_T$ is an idempotent in $\mathbb{C}[S_n]$.*

Proof First note that $V^\lambda := \mathbb{C}[S_n]\mathbf{E}_T = \mathbb{C}[S_n]\widehat{\mathbf{E}}_T$ since the two symmetrizers differ only by a scalar. Now, since $\mathbf{E}_T^2 = [\mathcal{H}_T V_T][\mathcal{H}_T V_T] = \mathcal{H}_T[V_T\mathcal{H}_T]V_T$, the fact that $\mathbf{E}_T^2 = \kappa \mathbf{E}_T$ for some $\kappa \in \mathbb{C}$ follows from Proposition 8.25.

It remains to determine the value of κ. For $a \in \mathbb{C}[S_n]$, let $R(\mathbf{E}_T)\colon \mathbb{C}[S_n] \to \mathbb{C}[S_n]$ be the linear map given by $R(\mathbf{E}_T)(a) := a\mathbf{E}_T$, that is, the "right-multiplication by $\mathbf{E}_T$" map. We will determine the value of κ by computing the trace of $R(\mathbf{E}_T)$ using two different bases for $\mathbb{C}[S_n]$.

For the first computation, we take the canonical basis $\{\sigma\}_{\sigma \in S_n}$. By Exercise 8.9,

$$\mathbf{E}_T = (1) + \sum \{\text{other terms not involving the identity}\}.$$

By Exercise 7.7, the identity term contributes $\operatorname{Dim}\mathbb{C}[S_n] = |S_n| = n!$ to the value of $\operatorname{tr} R(\mathbf{E}_T)$. All of the other non-identity terms in $\mathbf{E}_T$ contribute zero since none of the basis vectors in $\mathbb{C}[S_n]$ are fixed by any $\sigma \neq (1)$. Thus, by the properties of the trace, we have $\operatorname{tr} R(\mathbf{E}_T) = |S_n| = n!$.

For the second computation, let $\{\mathbf{v}_1, \ldots \mathbf{v}_d\}$ be a basis for V^λ, which we extend to a basis $\{\mathbf{v}_1, \ldots \mathbf{v}_d, \mathbf{w}_{d+1}, \ldots, \mathbf{w}_{n!}\}$ for $\mathbb{C}[S_n]$. If $\mathbf{v}_i, i = 1, \ldots, d$ is a basis vector for V^λ, then, by the definition of V^λ, we have that $\mathbf{v}_i = a_i\mathbf{E}_T$ for some $a_i \in \mathbb{C}[S_n]$. Therefore,

$$R(\mathbf{E}_T)\mathbf{v}_i = \mathbf{v}_i\mathbf{E}_T = (a_i\mathbf{E}_T)\mathbf{E}_T = a_i\mathbf{E}_T^2 = a_i\kappa\mathbf{E}_T = \kappa a_i\mathbf{E}_T = \kappa\mathbf{v}_i.$$

In other words, each basis vector $\mathbf{v}_i$ in V^λ is an eigenvector for $R(\mathbf{E}_T)$ with eigenvalue κ.

For the other basis vectors $\mathbf{w}_j$, $j = d + 1, \ldots, n!$, the definition of V^λ yields

$$R(\mathbf{E}_T)\mathbf{w}_j = \mathbf{w}_j\mathbf{E}_T \in V^\lambda,$$

and so $R(\mathbf{E}_T)\mathbf{w}_j$ is a linear combination of $\{\mathbf{v}_1, \ldots \mathbf{v}_d\}$. That is, the coefficient of $\mathbf{w}_j$ in $R(\mathbf{E}_T)\mathbf{w}_j$ is zero for $j = d + 1, \ldots, n!$, and so the jth diagonal entry in the matrix realization of $R(\mathbf{E}_T)$ is zero for $j = d+1, \ldots, n!$. With respect to this basis, the matrix realization of $R(\mathbf{E}_T)$ looks like this;

$$\widetilde{R(\mathbf{E}_T)} = \begin{pmatrix} \kappa & 0 & \ldots & 0 & & \\ 0 & \kappa & \vdots & 0 & \text{other} & \\ \vdots & & \ddots & & \text{stuff} & \\ & & & \kappa & & \\ 0 & & \ldots & 0 & & \\ \vdots & & & & \ddots & \\ 0 & & \ldots & & & 0 \end{pmatrix}.$$

So for this case, we have tr $R(\mathbf{E}_T) = d\kappa$. Consequently tr $R(\mathbf{E}_T) = n! = d\kappa$, and so $\kappa = \frac{n!}{d}$. $\square$

It is more efficient to prove the next result in full generality, then apply it to our situation.

Proposition 8.31 *Let $\mathfrak{e}$ be an idempotent in an algebra A, and set $A\mathfrak{e} := \{a\mathfrak{e} \mid a \in A\}$. If M is a left A-module, then $\mathrm{Hom}_A(A\mathfrak{e}, M) \cong \mathfrak{e}M$ as vector spaces.*

Proof Define $\psi : \mathfrak{e}M \to \mathrm{Hom}_A(A\mathfrak{e}, M)$ by

$$\psi(\mathfrak{e}m) := f_{\mathfrak{e}m} \in \mathrm{Hom}_A(A\mathfrak{e}, M), \text{ where } f_{\mathfrak{e}m}(x) := x\mathfrak{e}m, \quad x \in A\mathfrak{e}.$$

Define $\varphi : \mathrm{Hom}_A(A\mathfrak{e}, M) \to \mathfrak{e}M$ by

$$\varphi(f) := f(\mathfrak{e}).$$

Then ψ and φ are linear maps, with $\psi \circ \varphi = Id_{\mathrm{Hom}_A(A\mathfrak{e}, M)}$ and $\varphi \circ \psi = Id_{\mathfrak{e}M}$. Therefore ψ and φ are vector space isomorphisms. Here Id_V denotes the identity map on a vector space V. $\square$

Exercise* 8.32 Fill in the details in the proof of Proposition 8.31.

Remark 8.33 $\mathrm{Hom}_A(A\mathfrak{e}, M) \cong \mathfrak{e}M$ may not be equivalent *as left A-modules* because $\mathfrak{e}M$ need not be a left A-module. See Exercise 8.35 below.

Proof of Theorem 8.22 In order to apply Proposition 8.31, we can assume that $\lambda \geq \mu$ in the lexicographic order, and set

$$A = \mathbb{C}[S_n], \quad \mathfrak{e} = \widehat{\mathbf{E}}_\lambda, \quad \text{and } M = \mathbb{C}[S_n]\widehat{\mathbf{E}}_\mu.$$

Then Proposition 8.31 translates as

$$\mathrm{Hom}_{S_n}(V^\lambda, V^\mu) = \mathrm{Hom}_{\mathbb{C}[S_n]}(\mathbb{C}[S_n]\widehat{\mathbf{E}}_\lambda, \mathbb{C}[S_n]\widehat{\mathbf{E}}_\mu) \cong \widehat{\mathbf{E}}_\lambda \mathbb{C}[S_n]\widehat{\mathbf{E}}_\mu.$$

If $\lambda \neq \mu$, then by Proposition 8.29,

$$\mathrm{Hom}_{S_n}(V^\lambda, V^\mu) \cong \widehat{\mathbf{E}}_\lambda \mathbb{C}[S_n]\widehat{\mathbf{E}}_\mu = \frac{1}{\kappa_\lambda}\frac{1}{\kappa_\mu}\mathcal{H}_\lambda\Big[\mathcal{V}_\lambda \mathbb{C}[S_n]\mathcal{H}_\mu\Big]\mathcal{V}_\mu = 0,$$

so by Corollary 3.17, the representations V^λ and V^μ are inequivalent.

If $\lambda = \mu$, applying the definitions and Proposition 8.30 yields

$$\mathrm{Hom}_{S_n}(V^\lambda, V^\lambda) \cong \widehat{\mathbf{E}}_\lambda \mathbb{C}[S_n]\widehat{\mathbf{E}}_\lambda = \frac{1}{\kappa_\lambda^2}\mathcal{H}_\lambda\Big[\mathcal{V}_\lambda \mathbb{C}[S_n]\mathcal{H}_\lambda\Big]\mathcal{V}_\lambda.$$

Now for each $a \in \mathcal{V}_\lambda \mathbb{C}[S_n]\mathcal{H}_\lambda \subset \mathbb{C}[S_n]$, Proposition 8.25 yields

$$\mathcal{H}_\lambda a \mathcal{V}_\lambda = \ell(a)\mathbf{E}_\lambda, \quad \text{where } \ell \colon \mathbb{C}[S_n] \to \mathbb{C} \text{ is linear.}$$

Thus $\widehat{\mathbf{E}}_\lambda \mathbb{C}[S_n]\widehat{\mathbf{E}}_\lambda$ is spanned by $\widehat{\mathbf{E}}_\lambda$ and so is one-dimensional, provided we can show that the map ℓ is not identically zero on $\mathcal{V}_\lambda \mathbb{C}[S_n]\mathcal{H}_\lambda$.

Since the identity permutation $(1) \in \mathbb{C}[S_n]$, we have $\mathcal{V}_\lambda(1)\mathcal{H}_\lambda \in \mathcal{V}_\lambda \mathbb{C}[S_n]\mathcal{H}_\lambda$, and

$$\mathcal{H}_\lambda\Big[\mathcal{V}_\lambda(1)\mathcal{H}_\lambda\Big]\mathcal{V}_\lambda = \mathbf{E}_\lambda^2 = \kappa \mathbf{E}_\lambda \neq 0 \quad \text{by Proposition 8.30.}$$

It follows that V^λ is an irreducible representation of S_n by Corollary 3.18. Since every irreducible representation of S_n appears in $\mathbb{C}[S_n]$ (Remark 6.4), the collection $\{V^\lambda \mid \lambda \vdash n\}$ must be all of the irreducible representations of S_n, labeled by the partitions of n.

Note also that the number of integer partitions of n is equal to the number of conjugacy classes of S_n, and the results of Chap. 7 confirm that the collection $\{V^\lambda \mid \lambda \vdash n\}$ must be all of the irreducible representations of S_n. $\qquad \square$

Remark 8.34 Another proof that the collection $\{V^\lambda \mid \lambda \vdash n\}$ is a complete set of pair-wise inequivalent representations of S_n goes something like this: A projection $p \in \mathbb{C}[G]$ is a *minimal projection* if $0 \neq p = q + r$ can only hold for projections q and r if one of q or r is zero. A projection is *central* if it lies in the center of $\mathbb{C}[G]$. A *minimal central projection* is a projection that is minimal in the set of central projections (*i.e.*, both q and r must be central), not necessarily minimal in the set of all projections.

The Young symmetrizers are minimal central projections in the group algebra. Any two such projections are either equivalent or disjoint, and the space of such projections is in one-to-one correspondence with the irreducible representations of the group. See [Si] III.7 for details.

Exercise* 8.35 We will work through an example to demonstrate that if A is an algebra, $\mathfrak{e}$ an idempotent in A, and M a left A-module, then $\mathfrak{e}M$ need not be a left A-module. Note first that for $\mathfrak{e}M$ to be a left A-module we need $a.\mathfrak{e}m \in \mathfrak{e}M$ for all $a \subset \Lambda$ and $m \in M$.

Let $\mathfrak{e} = \frac{1}{3}[(1) + (1, 2) - (1, 3) - (1, 3, 2)]$ be the (normalized) Young symmetrizer from Example 8.4, and let $A=\mathbb{C}[S_3]=M$. Since $\{(1), (1, 2), \ldots, (1, 3, 2)\}$ is a basis for $M = \mathbb{C}[S_3]$, it follows that $\{\mathfrak{e}.(1), \mathfrak{e}.(1, 2), \ldots, \mathfrak{e}.(1, 3, 2)\}$ spans $\mathfrak{e}M$. Verify that (up to sign) there are only 3 distinct vectors in this set, and that any one of them is a linear combination of the other two. So suppose $\{\mathbf{v_1}, \mathbf{v_2}\}$ is a basis for $\mathfrak{e}M$. Now we can find an element $\sigma \in S_3$ such that $\sigma.\mathbf{v_1} \notin \mathfrak{e}M$, and so $\mathfrak{e}M$ is not a left A-module.

Exercise* 8.36 Verify the claim at the end of the proof of Proposition 7.18, that the identity element $(1) \in S_3$ can be written as a linear combination of vectors in $\mathbb{C}[S_3]$, one from each irreducible subrepresentation. Hint: Basis vectors for the trivial and alternating representations are obvious. Use the Young symmetrizers for the two standard Young tableaux corresponding to $(2, 1, 0) \vdash 3$ to obtain two copies of $\mathcal{W}$ in $\mathbb{C}[S_3]$.

8.5   More Representations

In Exercises 8.15–8.20 we used the Young symmetrizers to obtain irreducible representations of S_n in polynomial spaces. This procedure generalizes, and provides a way to produce irreducible representations of S_n in any representation of S_n.

Proposition 8.37 *Let G be a group, let (ρ, V) be a representation of the group algebra $\mathbb{C}[G]$, and let $\mathbf{v} \in V$. The map $\phi_{\mathbf{v}} : \mathbb{C}[G] \to V$ given by $\phi_{\mathbf{v}}(a) := \rho(a)\mathbf{v}$ intertwines the left regular representation L of G on $\mathbb{C}[G]$ and the representation ρ of $G \subset \mathbb{C}[G]$ on V.*

Proof

$$\rho(g)[\phi_{\mathbf{v}}(a)] = \rho(g)[\rho(a)\mathbf{v}] = \rho(ga)\mathbf{v} = \rho(L(g)a)\mathbf{v} = \phi_{\mathbf{v}}(L(g)a).$$

$\square$

Since the set $\{a\widehat{\mathbf{E}}_T \mid a \in \mathbb{C}[S_n]\}$ spans the irreducible representation V^λ in $\mathbb{C}[S_n]$, it follows that the set $\{\phi_{\mathbf{v}}(a\widehat{\mathbf{E}}_T) \mid a \in \mathbb{C}[S_n]\} = \{a\widehat{\mathbf{E}}_T.\mathbf{v} \mid a \in \mathbb{C}[S_n]\}$ spans an irreducible representation in V for any $\mathbf{v} \in V$.

As an example, here is a slight variation of Exercise 8.20 that uses Young symmetrizers as projections in $\mathcal{P}_1$. For the standard Young tableau given by

$$T = \begin{array}{|c|c|} \hline 1 & 2 \\ \hline 3 \\ \cline{1-1} \end{array}$$

we set

$$\widehat{\mathbf{E}}_1 := \widehat{\mathbf{E}}_T = \frac{1}{3}[(\mathbf{1}) + (\mathbf{1},\mathbf{2})][(\mathbf{1}) - (\mathbf{1},\mathbf{3})] = (\mathbf{1}) + (\mathbf{1},\mathbf{2}) - (\mathbf{1},\mathbf{3}) - (\mathbf{1},\mathbf{3},\mathbf{2})$$

and set

$$\widehat{\mathbf{E}}_2 := (1,3).\widehat{\mathbf{E}}_1 = \frac{1}{3}[-(\mathbf{1}) + (\mathbf{1},\mathbf{3}) - (\mathbf{2},\mathbf{3}) + (\mathbf{1},\mathbf{2},\mathbf{3})],$$

which gives us two linearly independent vectors that span a copy of $\mathcal{W}$ in $\mathbb{C}[S_3]$. Applying Proposition 8.37, we have

$$\phi_{\mathbf{x}}(\widehat{\mathbf{E}}_1) = \widehat{\mathbf{E}}_1.x = \frac{1}{3}(x + y - 2z),$$

and

$$\phi_{\mathbf{x}}(\widehat{\mathbf{E}}_2) = \widehat{\mathbf{E}}_2.x = \frac{1}{3}(-2x + y + z),$$

which are linearly independent vectors that span a copy of $\mathcal{W}$ in $\mathcal{P}_1(x, y, z)$. Try this with $\phi_{\mathbf{y}}$ and $\phi_{\mathbf{z}}$.

Remark 8.38 For more advanced readers, the use of Young symmetrizers as projections works for any representation of S_n, including tensor product spaces, where, for example,

$$(1,2).\mathbf{v}_1 \otimes \mathbf{v}_2 \otimes \mathbf{v}_3 = \mathbf{v}_2 \otimes \mathbf{v}_1 \otimes \mathbf{v}_3.$$

There is a refinement of sorts for the Young symmetrizers called *semi-normal idempotents*. These are defined inductively, and provide an inductive basis (see Remark 9.4) for irreducible representations in $\mathbb{C}[S_n]$, and so for any representation onto which they project. See [JK], Section 3.2. For an application, see [MS].

Exercise* 8.39 Here we show directly that $\mathbf{E}_T := \mathcal{V}_T \mathcal{H}_T$, and the "backwards" definition that we call $\mathbf{E}_T^* := \mathcal{H}_T \mathcal{V}_T$ in Remark 8.6 result in equivalent left S_n-modules.

Let T be a Young tableau with shape λ, let $V^\lambda = \mathbb{C}[S_n]\mathbf{E}_T$, and define $V^{\lambda^*} := \mathbb{C}[S_n]\mathbf{E}_T^*$. Recall that $\mathbf{E}_T^2 = \kappa \mathbf{E}_T$ for some positive $\kappa \in \mathbb{C}$, and then for $x \in V^\lambda$ define

$$\Phi \colon V^\lambda \to V^{\lambda^*} \text{ by } \Phi(x) := x\left[\frac{1}{\sqrt{\kappa}}\mathcal{H}_T\right].$$

In other word, Φ is the "right multiplication by $\left[\frac{1}{\sqrt{\kappa}}\mathcal{H}_T.\right]$ map." Similarly, for $y \in V^{\lambda^*}$ define

$$\Psi \colon V^{\lambda^*} \to V^\lambda \text{ by } \Psi(y) := y\left[\frac{1}{\sqrt{\kappa}}\mathcal{V}_T.\right]$$

(1) Verify that the maps Φ and Ψ are S_n-linear maps. Also check that $\Phi(x)$ actually lies in V^{λ^*}, and that $\Psi(y)$ actually lies in V^λ.
(2) Show that $\Psi \circ \Phi(x) = x$, and that $\Phi \circ \Psi(y) = y$. Conclude that Φ and Ψ are S_n-isomorphisms, and thus V^λ and V^{λ^*} are equivalent left S_n-modules.

8.6   Hints and Additional Comments

Exercise 8.10 Note first that the action of each $\sigma \in S_n$ on T just relabels the boxes in the Young diagram for T. Thus if $i, j \in \{1, \ldots n\}$ are in the same row (column) of T, then $\sigma.i$ and $\sigma.j$ must also be in the same row (column) of $\sigma.T$.

Now let $h \in H_T$. We want to show that $\sigma h \sigma^{-1} \in H_{\sigma.T}$, and therefore $\sigma H_T \sigma^{-1} \subseteq H_{\sigma.T}$. So suppose that $i \in \{1, \ldots n\}$ is an entry in T. Then $\sigma.i$ is an entry in $\sigma.T$, and

$$\sigma h \sigma^{-1}.(\sigma.i) = \sigma h \sigma^{-1}\sigma.i = \sigma h.i .$$

Since $h \in H_T$, the entries i and $h(i)$ are in the same row of T. By the above "note first" remark, $\sigma.i$ and $\sigma h.i$ must be in the same row of $\sigma.T$, and therefore $\sigma x \sigma^{-1} \in H_{\sigma.T}$. The opposite inclusion, that $H_{\sigma.T} \subseteq \sigma H_T \sigma^{-1}$, follows by the same argument since $H_T = H_{\sigma^{-1}.(\sigma.T)}$. Obviously the same process shows that V_T and $V_{\sigma.T}$ are conjugate, and the conclusion follows from the definition of $\mathbf{E}_T$.

Exercise 8.11 Let G_s and $G_{s'}$ denote the stabilizers of s and s' respectively. If s' is in the same G-orbit as s, then $s' = x.s$ for some $x \in G$. Now suppose $g \in G_s$. Then

$$
\begin{aligned}
xgx^{-1}.s' &= xg.s \quad \text{(since } x^{-1}.s' = s) \\
&= x.s \quad \text{(since } g.s = s) \\
&= s'.
\end{aligned}
$$

Therefore $xG_sx^{-1} \subset G_{s'}$. A similar argument shows the opposite inclusion.

Exercise 8.14 By Exercise 8.12, if $v \in V_T$, then $vV_T = \mathrm{sgn}(v)V_T$. So

$$
V_T^2 = V_T V_T = \left[\sum_{v \in V_T} \mathrm{sgn}(v)v \right] V_T = \sum_{v \in V_T} \mathrm{sgn}(v)\,[vV_T] = \sum_{v \in V_T} \mathrm{sgn}(v)^2 V_T = |V_T| V_T.
$$

Exercises 8.15–8.20 We hope you concluded that the partition $(2, 1, 0) \vdash 3$ corresponds to the standard representation W_3, and that there can be two distinct copies of W_3 corresponding to the two standard Young tableaux associated to the partition $(2, 1, 0)$. But when W_3 appears only once, as in P_1, they project to the same subspace or onto the zero subspace. Also, the partition $(3, 0, 0)$ corresponds to the identity representation I_3, and that the partition $(1, 1, 1)$ corresponds to the alternating representation A_3, each of which have only one possible standard Young tableau associated to them. Note that applying the Young symmetrizer corresponding to the partition $(1, 1, 1)$ to the space P_1 yields only the zero subspace since the alternating representation does not appear in P_1. Applying the Young symmetrizer for $(1, 1, 1)$ to the space $P_{(2,1,0)}$ gives the alternating representation A_3 (up to a scalar multiple) that we obtained in Sect. 5.3.

Remark 8.23 What is a generating function? A simple example that everyone should be familiar with is the generating function for the binomial coefficients;

$$
\binom{n}{k} = \frac{n!}{k!(n-k)!},
$$

whose value gives the number of ways to choose k unordered items from n possible choices. The binomial theorem says that

$$
(x + 1)^n = \sum_{k=0}^{n} \binom{n}{k} x^{n-k},
$$

so the binomial coefficients are the coefficients of the the polynomial $(x+1)^n$, which is the generating function.

More generally, a generating function $f(x)$ is a formal power series

$$f(x) = \sum_{n=0}^{\infty} a_n x^n$$

whose coefficients give the sequence $\{a_0, a_1, \ldots\}$. For example (see Remark 2.2),

$$\frac{1}{1-x} = \sum_{n=0}^{\infty} x^n,$$

so $\frac{1}{1-x}$ is the generating function for the sequence $\{1, 1, 1, \ldots\}$. Similarly,

$$\frac{1}{(1-x)^2} = \sum_{n=0}^{\infty} (n+1)x^n,$$

so $\frac{1}{(1-x)^2}$ is the generating function for the sequence $\{1, 2, 3, \ldots\}$ (as if we need one).

As a more substantial example, let $P(n)$ denote the *partition function*, that is, the number of partitions of a non-negative integer n. Then

$$\sum_{n=0}^{\infty} P(n)x^n = \prod_{j=1}^{\infty} \frac{1}{1-x^j} = (1+x+x^2+x^3+\cdots)(1+x^2+x^4+x^6+\cdots)(1+x^3+x^6+\cdots)\cdots.$$

Unlike the case for binomial coefficients, $P(n)$ has no (known) closed form expression. You should work out the first few terms.

Exercise 8.24

(1) $\mathcal{H}_T(1, 3, 2)\mathcal{V}_T = -\mathcal{H}_T\mathcal{V}_t$.
(2) $\mathcal{H}_T(2, 3)\mathcal{V}_T = 0$.
(3) $\mathcal{H}_T(1, 2, 3)\mathcal{V}_T = 0$.

Exercise 8.26   Check that

$$(4, 5)(1, 3).T = \begin{array}{|c|c|c|}
\hline
3 & 2 & 1 \\
\hline
5 & 4 & 6 \\
\hline
7 \\
\cline{1-1}
\end{array} = (5, 7)(2, 4).T',$$

or equivalently, that

$$(2, 4)(5, 7)(4, 5)(1, 3).T = T'$$

for $(4, 5)(1, 3) \in H_T$ and $(5, 7)(2, 4) \in V_{T'}$.

Since $h \in H_T$ permutes only the rows of T, it should be readily apparent that T and $h.T$ have the same horizontal subgroups (or use Exercise 8.10).

Referring to the statement of the problem, we see that

$$(2,4)(5,7)(4,5)(1,3)T = (4,5)(1,3)(2,5)(4,7).T = T',$$

and hence

$$(v_2'v_1')^{-1}h_2h_1 = (2,4)(5,7)(4,5)(1,3) = (4,5)(1,3)(2,5)(4,7) \in H_T V_T.$$

Exercise 8.32 It is a useful exercise to verify everything in sight:

(1) For $em \in eM$, x and $y \in Ae$, r and $s \in \mathbb{C}$, and for $a \in A$, define $f_{em}(x) := xem$, (the "right multiplication by em" map). Then the map

$$\psi : eM \to \mathrm{Hom}_A(Ae, M), \ \text{ given by } \psi(em) := f_{em} \in \mathrm{Hom}_A(Ae, M),$$

is a linear map between the given vector spaces.

(a) The image of f_{em} lies in M because M is a left A-module.

(b) The map $f_{em} \in \mathrm{Hom}_A(Ae, M)$ since:

    (i) $f_{em}(rx + sy) = (rx + sy)(em) = rx(em) + sy(em) = rf_{em}(x) + sf_{em}(y)$,

       so f_{em} is linear.

    (ii) $f_{em}(ax) = (ax)(em) = ax(em) = af_{em}(x)$, so f_{em} is an A-map.

(c) The map ψ itself is linear since for $em, en \in eM$, $x \in Ae$, and for $r, s \in \mathbb{C}$, we have;

$$\begin{aligned}
\psi[rem + sen](x) &= f_{[rem+sen]}(x) \\
&= x[rem + sen] \\
&= xrem + xsen \\
&= rxem + sxen \\
&= rf_{em}(x) + sf_{en}(x) \\
&= r\psi(em) + s\psi(en).
\end{aligned}$$

(2) For $f, g \in \mathrm{Hom}_A(Ae, M)$, r and $s \in \mathbb{C}$, and $a \in A$, the "evaluation at e" map,

$$\varphi : \mathrm{Hom}_A(Ae, M) \to eM, \ \text{ defined by } \varphi(f) := f(e),$$

is also a linear map between the given vector spaces.

(a) Since A has an identity 1_A, the idempotent $e = 1_A e \in Ae$.

(b) Since f is a left A-map, $e \in A$, and $f(e) \in M$, we have $f(e) = f(e^2) = ef(e) \in eM$.

(c) The map φ itself is linear since $\varphi(rf + sg) = [rf + sg](e) = rf(e) + sg(e) = r\varphi(f) + s\varphi(g)$.

(3) The composition $\psi \circ \varphi = Id_{Hom_A(A\mathfrak{e}, M)}$ since for any $a\mathfrak{e} \in A\mathfrak{e}$, and for all $f \in Hom_A(A\mathfrak{e}, M)$, we have

$$
\begin{aligned}
[\psi \circ \varphi(f)](a\mathfrak{e}) &= \psi[f(\mathfrak{e})](a\mathfrak{e}) \quad (\text{since } \varphi(f) = f(\mathfrak{e})) \\
&= a\mathfrak{e}[f(\mathfrak{e})] \quad (\psi[f(\mathfrak{e})] \text{ is right multiplication by } f(\mathfrak{e})) \\
&= f(a\mathfrak{e}^2) \quad (f \text{ is an } A\text{-map, and } a\mathfrak{e} \in A) \\
&= f(a\mathfrak{e}). \quad (\mathfrak{e}^2 = \mathfrak{e})
\end{aligned}
$$

That is, $\psi \circ \varphi(f) = f$.

(4) Similarly, $\varphi \circ \psi = Id_{\mathfrak{e}M}$ since for all $\mathfrak{e}m \in \mathfrak{e}M$,

$$[\varphi \circ \psi](\mathfrak{e}m) = \varphi(f_{\mathfrak{e}m}) = f_{\mathfrak{e}m}(\mathfrak{e}) = \mathfrak{e}^2 m = \mathfrak{e}m.$$

Exercise 8.35 When we perform some computations, we find that $\mathfrak{e}M = \mathfrak{e}\mathbb{C}[S_3]$ is spanned by the three vectors $\mathbf{v_1} := \mathfrak{e}.(1)$, $\mathbf{v_2} := \mathfrak{e}.(1, 2)$, and $\mathbf{v_3} := \mathfrak{e}.(2, 3)$. (Note that $\mathfrak{e}.(1, 3) = -\mathbf{v_1}$, $\mathfrak{e}.(1, 2, 3) = -\mathbf{v_2}$ and $\mathfrak{e}.(1, 3, 2) = -\mathbf{v_3}$.)

Also note that $\mathbf{v_3} = \mathbf{v_1} - \mathbf{v_2}$, so we can choose, say, $\{\mathbf{v_1}, \mathbf{v_2}\}$, as a basis for $\mathfrak{e}M$. Then verify that $(2, 3).\mathbf{v_2}$ cannot be a linear combination of $\{\mathbf{v_1}, \mathbf{v_2}\}$, and so $(2, 3).\mathbf{v_2} \notin \mathfrak{e}M$. Therefore $\mathfrak{e}M$ cannot be a left $\mathbb{C}[S_3]$-module.

Exercise 8.36 The trivial representation is spanned by

$$\mathbf{I} := (\mathbf{1}) + (\mathbf{1, 2}) + (\mathbf{1, 3}) + (\mathbf{2, 3}) + (\mathbf{1, 2, 3}) + (\mathbf{1, 3, 2}) = \sum_{\sigma \in S_n} \sigma.$$

The alternating representation is spanned by

$$\mathbf{A} := (\mathbf{1}) - (\mathbf{1, 2}) - (\mathbf{1, 3}) - (\mathbf{2, 3}) + (\mathbf{1, 2, 3}) + (\mathbf{1, 3, 2}) = \sum_{\sigma \in S_n} \text{sgn}(\sigma)\sigma.$$

One copy of $\mathcal{W}$, corresponding to the standard Young tableau

$$T_1 = \begin{array}{|c|c|} \hline 1 & 2 \\ \hline 3 \\ \cline{1-1} \end{array},$$

is spanned by

$$\mathbf{E_1} := (\mathbf{1}) + (\mathbf{1, 2}) - (\mathbf{1, 3}) - (\mathbf{1, 3, 2}) \text{ and } (2, 3).\mathbf{E_1} := -(\mathbf{1, 2}) + (\mathbf{2, 3}) - (\mathbf{1, 2, 3}) + (\mathbf{1, 3, 2}).$$

The other copy of $\mathcal{W}$, corresponding to the standard Young tableau

$$T_2 = \begin{array}{|c|c|} \hline 1 & 3 \\ \hline 2 \\ \cline{1-1} \end{array},$$

is spanned by

$$\mathbf{E}_2 := (1) - (1,2) + (1,3) - (1,2,3) \text{ and } (2,3).\mathbf{E}_2 := -(1,3) + (2,3) + (1,2,3) - (1,3,2).$$

Solving the system

$$(1) = r_1\mathbf{I} + r_2\mathbf{A} + r_3\mathbf{E}_1 + r_4(2,3).\mathbf{E}_1 + r_5\mathbf{E}_2 + r_6(2,3).\mathbf{E}_2$$

for the scalars $r_1, \ldots, r_6$ yields

$$r_1 = r_2 = 1/6, \quad r_3 = r_5 = 1/3, \quad \text{and } r_4 = r_6 = 0.$$

Here we have used the fact that, for example, $\mathbf{E}_1$ generates a copy of $\mathcal{W}$, so for all $\sigma \in S_3$, the element $\sigma.\mathbf{E}_1$ lies in that same copy of $\mathcal{W}$. Just be sure to choose σ so that $\mathbf{E}_1$ and $\sigma.\mathbf{E}_1$ are linearly independent.

Exercise 8.39 Verifying that Φ and Ψ are S_n-linear is straightforward. For example,

$$\Phi(\sigma.x) = \sigma.x \left[\frac{1}{\sqrt{\kappa}}\mathcal{H}_T\right] = \sigma. \left[x\frac{1}{\sqrt{\kappa}}\mathcal{H}_T\right] = \sigma.\Phi(x).$$

If $x \in V^\lambda$, then $x = a\mathbf{E}_T$ for some $a \in \mathbb{C}[S_n]$. Therefore,

$$\begin{aligned}
\Phi(x) &= x\left[\frac{1}{\sqrt{\kappa}}\mathcal{H}_T\right] \\
&= a\mathbf{E}_T\left[\frac{1}{\sqrt{\kappa}}\mathcal{H}_T\right] \\
&= a\mathcal{H}_T\mathcal{V}_T\left[\frac{1}{\sqrt{\kappa}}\mathcal{H}_T\right] \\
&= \frac{1}{\sqrt{\kappa}}a\mathcal{H}_T[\mathcal{V}_T\mathcal{H}_T] \\
&= \frac{1}{\sqrt{\kappa}}a\mathcal{H}_T\mathbf{E}_T^* \in V^{\lambda^*}.
\end{aligned}$$

Using the previous result, we have

$$\begin{aligned}
\Psi \circ \Phi(x) &= \Psi\left(\frac{1}{\sqrt{\kappa}}a\mathcal{H}_T[\mathcal{V}_T\mathcal{H}_T]\right) \\
&= \left(\frac{1}{\sqrt{\kappa}}a\mathcal{H}_T[\mathcal{V}_T\mathcal{H}_T]\right)\frac{1}{\sqrt{\kappa}}\mathcal{V}_T \\
&= \frac{1}{\kappa}a\mathcal{H}_T\mathcal{V}_T\mathcal{H}_T\mathcal{V}_T \\
&= \frac{1}{\kappa}a\mathbf{E}_T^2 \\
&= \frac{1}{\kappa}a\kappa\mathbf{E}_T \\
&= a\mathbf{E}_T \\
&= x.
\end{aligned}$$

Similarly, $\Phi \circ \Psi(y) = y$. By the way, if the field of scalars has positive characteristic, we could have $\kappa = 0$ and this argument would fail.

Cosets, Restricted and Induced Representations

9

Given a representation of a group, obtaining a representation of one of its subgroups is straightforward. The reverse problem is more interesting; given a representation of a subgroup of a group, can we obtain a representation of the group itself? We will present two constructions that result in the "best" representation of a group given a representation of one of its subgroups, and show that in our case they are equivalent. Along the way we will need to develop more group theory.

Many of the results in this section are stated and proven for the case of finite groups. Serious readers are encouraged to investigate their generalizations to other groups.

9.1   Restriction

Definition 9.1 Given a representation of a group G on a vector space V and a subgroup H of G, we obtain a representation W of H by simply restricting the G-action to H. In this case we write $W = \mathrm{Res}_H^G V$, or just $W = \mathrm{Res}\, V$ if there is no ambiguity about H and G. If we need to specify the group actions, we write $\rho = \mathrm{Res}_H^G \pi$ for ρ a representation of H and π a representation of G. Similarly we will use the notation $\chi_\rho = \chi_{\mathrm{Res}_H^G \pi}$ or $\chi_W = \chi_{\mathrm{Res}_V}$ for *restricted characters*, that is, characters of restricted representations. The notation $W = V \downarrow_H^G$ and $\rho = \pi \downarrow_H^G$ is also common.

Example 9.2 The trivial representation of S_3 on $\mathcal{I}_3 = \langle\!\langle \mathbf{x} + \mathbf{y} + \mathbf{z} \rangle\!\rangle \subset \mathcal{P}_1(x, y, z)$ restricts to the trivial representation $\mathcal{I}_2$ of S_2 since the basis vector $\mathbf{x} + \mathbf{y} + \mathbf{z}$ remains invariant if we permute only the $\mathbf{x}$ and $\mathbf{y}$.

Example 9.3 How does the standard representation of S_3, for example

$$W_3 = \{r\mathbf{x} + s\mathbf{y} + t\mathbf{z} \mid r + s + t = 0\} \subset \mathcal{P}_1(x, y, z),$$

© The Author(s), under exclusive license to Springer Nature Switzerland AG 2022
R. M. Howe, *An Invitation to Representation Theory*, SUMS Readings,
https://doi.org/10.1007/978-3-030-98025-2_9

restrict to S_2? The best way to see this is by the clever choice of a basis for $\mathcal{W}_3$. Clearly $\{\mathbf{x} - \mathbf{y}, \mathbf{x} + \mathbf{y} - 2\mathbf{z}\}$ is a basis for $\mathcal{W}_3$, and when we restrict the group action to the standard embedding of S_2 in S_3, we see that the vector $\mathbf{x} - \mathbf{y}$ changes sign when the variables $\mathbf{x}$ and $\mathbf{y}$ are transposed, while the vector $\mathbf{x} + \mathbf{y} - 2\mathbf{z}$ remains invariant. Thus the S_3-irreducible space $\mathcal{W}_3$ decomposes into $\mathcal{I}_2 \oplus \mathcal{A}_2$ when the action is restricted to the subgroup S_2. Had we used another basis for $\mathcal{W}_3$, the decomposition of $\operatorname{Res}_{S_2}^{S_3} \mathcal{W}_3$ would not have been so obvious.

Note that restrictions of irreducible representations need not be irreducible.

Remark 9.4 A basis like the nifty one above, that yields a nicely-behaved basis upon restriction to a subgroup, can be very useful and appears often in representation theory. Such a basis is an example of an *inductive basis*, a *Gelfand-Tsetlin* basis, or a *Gelfand-Zetlin* (abbreviated *G-Z*) basis. Occasionally, we see the term *Young basis* in the context of the symmetric group.

Exercise* 9.5 Let $\mathcal{W}_4$ be the three-dimensional standard representation of S_4 in $\mathcal{P}_1(x, y, z, w)$ defined by

$$\mathcal{W}_4 := \{r_1\mathbf{x} + r_2\mathbf{y} + r_3\mathbf{z} + r_4\mathbf{w} \mid r_1 + r_2 + r_3 + r_4 = 0\}.$$

How does this representation decompose when restricted to S_3? Hint: Mimic the discussion from Example 2.29 to show that $\mathcal{W}_4$ is irreducible and three dimensional, then find a Young basis. Extra hint: $\mathbf{x} - \mathbf{y}$ and $\mathbf{x} + \mathbf{y} - 2\mathbf{z}$ are two linearly independent vectors in $\mathcal{W}_4$. What is a good choice for the third basis vector?

9.2   Quotient Spaces

Let's recall the notion of *quotient spaces* from linear algebra. Given a vector space V with subspace W, for each $\mathbf{v} \in V$ the set

$$\mathbf{v} + W := \{\mathbf{v} + \mathbf{w} \mid \mathbf{w} \in W\}$$

is called the *affine space* through $\mathbf{v}$ parallel to W, or the *left coset* of W containing $\mathbf{v}$. The set of all left cosets is called the *quotient space* of V modulo W ("V mod W" for short), written V/W, and is itself a vector space.

As a simple example, let $V = \mathbb{R}^2$ and let $W = \{(0, r) \mid r \in \mathbb{R}\}$ be the y-axis. Then

$$\mathbf{v} + W = \{(x, y) + (0, r) \mid x, y, r \in \mathbb{R}\} = \{(x, y + r) \mid x, y, r \in \mathbb{R}\} = (x, 0) + W.$$

That is, the quotient space "looks like" the x-axis. Note that $(1, 0) + W = (1, 1) + W = (1, 3) + W = \cdots$. The following exercise is a good review, and is useful in its own right.

Exercise* 9.6 Let $V = U \oplus W$ be a finite-dimensional vector space. For $\mathbf{v} \in V$, define the left coset $\mathbf{v} + W$ as above, and denote the collection of these left cosets by V/W.

(1) Show that $\mathbf{v} + W = \mathbf{v}' + W$ if and only if $\mathbf{v} - \mathbf{v}' \in W$, and therefore $\mathbf{v} = \mathbf{v}' + \mathbf{w}$ for some $\mathbf{w} \in W$.
(2) Define addition and scalar multiplication on cosets by

$$[\mathbf{v} + W] + [\mathbf{v}' + W] := (\mathbf{v} + \mathbf{v}') + W, \quad \text{and} \quad r[\mathbf{v} + W] := r\mathbf{v} + W.$$

Show that these operations are well-defined using part (1). Conclude that V/W is a vector space, with $W = \mathbf{0} + W$ as the additive identity.
(3) Show that, for any $\mathbf{v}, \mathbf{v}' \in V$, either $\mathbf{v} + W = \mathbf{v}' + W$, or $\mathbf{v} + W \cap \mathbf{v}' + W = \{0\}$.
(4) Let $\{\mathbf{u}_1, \ldots, \mathbf{u}_k\}$ be a basis for the subspace U of V. Show that $\{\mathbf{u}_1 + W, \ldots, \mathbf{u}_k + W\}$ is a basis for V/W. Conclude that $\mathrm{Dim}(V/W) = \mathrm{Dim}\,V - \mathrm{Dim}\,W$, and consequently that V/W and U are isomorphic as vector spaces.
(5) Show that the map $\phi: V \to V/W$ given by $\phi(\mathbf{v}) = \mathbf{v} + W$ is a linear map onto V/W.
(6) Let $V = U \oplus W$ be a representation of a group G, where U and W are invariant subspaces. Define an action of G on V/W by $g.[\mathbf{v} + W] = g.\mathbf{v} + W$. Show that this action is well-defined. Show that the map ϕ in part (5) is a G-map, and conclude that U and V/W are equivalent representations of G.

Exercise* 9.7 Let V and W be finite-dimensional representations of G, and let $\phi: V \to W$ be a G-map. If $K = \mathrm{Ker}\,\phi$, show that $\mathrm{Im}\,\phi \cong V/K$. Conclude that $V = K \oplus U$, where $U \cong \mathrm{Im}\,\phi$. Hint: Define $\widehat{\phi} : V/K \to W$ by $\widehat{\phi}(\mathbf{v} + K) = \phi(\mathbf{v})$, and show that $\widehat{\phi}$ is a G-isomorphism.

Exercise* 9.8 Let G be a finite group. Show that every irreducible representation of a group G appears in $\mathbb{C}[G]$. Hint: Let V be an irreducible representation of G, and for $0 \neq \mathbf{v} \in V$, define $\phi : \mathbb{C}[G] \to V$ by $\phi(a) := a.\mathbf{v}$. Then use Exercise 9.7.

Corollary 9.9 *The collection* $\{V^\lambda \mid \lambda \vdash n\}$ *is a complete set of irreducible representations of* S_n, *as we have previously noted.*

9.3 Cosets

There are constructions analogous to quotient spaces for other mathematical objects, and for the case of groups, these constructions are called cosets. Cosets are an important tool for the study of groups, and we will them need for our discussion of induced representations.

Definition 9.10 Let H be a subgroup of a group G, and let $g \in G$. Define the *left coset of H containing g*, denoted gH, as $gH := \{gh \mid h \in H\}$. The element $g \in G$ is called a *coset representative* of gH. The set of all left cosets of H in G is written G/H, and pronounced "*G* modulo *H*" or "*G* mod *H*."

Remark 9.11 We can similarly define the *right coset containing g*, as $Hg := \{hg \mid h \in H\}$, and the *double coset containing g and g'*, as $gHg' := \{ghg' \mid h \in H\}$.

Only in the special case of when H is a so-called *normal* subgroup of G can we define a product of cosets, and thereby give G/H a group structure. Readers who are serious about modern algebra should look up "normal subgroups," and check that the multiplication given by $(aH)(bH) = abH$ is well-defined. Also check that the alternating group A_n is the only proper normal subgroup of S_n for $n \geq 5$.

Example 9.12 Let $G = GL(2, \mathbb{R})$, let $H = SL(2, \mathbb{R})$, and let $g = \left(\begin{smallmatrix} 2 & 0 \\ 0 & 1 \end{smallmatrix}\right) \in G$. How can we more completely describe the coset gH?

If $x \in gH$, then $x = gh$ for some $h \in H$. Using the definition of H (those matrices with determinant equal to 1) and the properties of determinants, we have $\text{Det } x = \text{Det}(gh) = \text{Det}(g)\,\text{Det}(h) = 2$.

Conversely, suppose some $x = \left(\begin{smallmatrix} a & b \\ c & d \end{smallmatrix}\right) \in G$ has determinant equal to 2. Then we can factor x as

$$x = \begin{pmatrix} a & b \\ c & d \end{pmatrix} = \begin{pmatrix} 2 & 0 \\ 0 & 1 \end{pmatrix} \begin{pmatrix} \frac{a}{2} & \frac{b}{2} \\ c & d \end{pmatrix}.$$

After a quick calculation, we see that the right-hand factor has determinant equal to 1 and so lies in H. Thus $x = gh$ for some $h \in H$, and we conclude that gH consists of all matrices in $GL(2, \mathbb{R})$ with determinant equal to 2. Note that there can be lots of coset representatives. For example,

$$gH = \begin{pmatrix} 2 & 0 \\ 0 & 1 \end{pmatrix} H = \begin{pmatrix} 0 & -2 \\ 1 & 0 \end{pmatrix} H = \begin{pmatrix} 2 & 0 \\ 1 & 1 \end{pmatrix} H = \begin{pmatrix} \sqrt{2} & 0 \\ 1 & \sqrt{2} \end{pmatrix} H = \dots.$$

Example 9.13 Let $G = (\mathbb{Z}, +)$, let $H = (3\mathbb{Z}, +)$, and let $g = 2 \in G$. Since the group is additive, the left coset is written $g + H = 2 + H = \{2 + 3k \mid k \in \mathbb{Z}\} = \{\dots, -4, -1, 2, 5, 8, \dots\}$. Again, note that $2 + H = 5 + H = \dots$.

You should try and prove the results in Proposition 9.14 below before looking them up. Hint: Exercise 9.6. You should also verify these properties for some of the above examples.

Proposition 9.14 *Let H be a subgroup of a group G, and let $a, b \in G$. Then*

(1) $a \in aH$,
(2) $aH = H$ if and only if $a \in H$,
(3) $aH = bH$ if and only if $a \in bH$,

(4) either $aH = bH$ or $aH \cap bH$ is empty,
(5) $aH = bH$ if and only if $a^{-1}b \in H$,
(6) $G = \biguplus_{g \in G} gH$ (disjoint union),
(7) $|aH| = |bH|$.

The above properties imply that the relation "$a \sim b$ if $aH = bH$" is an equivalence relation on G, and it follows that a group G is the disjoint union of its distinct left H-cosets, each of which have the same number of elements.[1] Consequently, if G is finite, $|G|$ is divisible by $|H|$, and $|G/H|$, the number of left cosets, is equal to $|G|/|H|$. This result is known as *Lagrange's Theorem*. The number of left cosets of H in G is called the *index of H in G*, written $[G : H]$.

Definition 9.15 A selection of exactly one coset representative from each coset of H is called a *transversal to H in G*, a *transversal of G/H*, or occasionally, a *system of representatives* of G/H. In the case where $[G : H]$ is finite, and where $\{g_1, g_2, \ldots, g_k\}$ is a transversal of G/H, this means that $G/H = \{g_1 H, g_2 H, \ldots, g_k H\}$ is a complete set of disjoint left cosets of H in G. It is customary and convenient to set $g_1 = e$, the identity in G.

Exercise* 9.16 Referring to the above examples, write down a transversal for H in G, where:

(1) $G = GL(2, \mathbb{R})$, and $H = SL(2, \mathbb{R})$;
(2) $G = (\mathbb{Z}, +)$, and $H = (3\,\mathbb{Z}, +)$.

Exercise* 9.17 Generalize the construction of Example 9.13 for the subgroup $H = k\mathbb{Z}$ for some fixed integer k. Write down the obvious set of coset representatives for each coset $a + k\mathbb{Z}$.

Exercise* 9.18 Let $G = (\mathbb{C}^*, \times)$, the multiplicative group of non-zero complex numbers, and let H be the subgroup $\{a + bi \mid a^2 + b^2 = 1\}$ from Exercise 1.47. Choose an element, say $g = 1 + 2i \in G$. Describe the coset gH geometrically. How does G/H partition G into disjoint subsets? What is a transversal for G/H?

Exercise* 9.19 Let $G = (\mathbb{R}^2, +)$, and let H be the subgroup $\{(x, 2x) \mid x \in \mathbb{R}\}$ from Exercise 1.44. Choose an element, say $g = $ the point $(1, 3)$ in the plane. Describe the left coset $g + H$ geometrically. How does G/H partition G into disjoint subsets? What is a transversal for G/H?

---

[1] A collection of non-empty disjoint subsets of a set S whose union is S is called a *partition* of S. This is a *set partition*, as opposed to the *integer partitions* encountered previously. Any equivalence relation on a set partitions the set into equivalence classes. Note the use of the word "partition" as both a noun and a verb.

Exercise 9.20 Write down all the subgroups of S_3, and then list all the elements in each coset of each subgroup. Write down a transversal for each case.

Exercise 9.21 Write out the cosets in S_4 for the following subgroups H of S_4 and determine a set of coset representatives.

(1) H = the alternating subgroup A_4.
(2) H = the dihedral group D_4 from Exercise 1.53. Note that the 2-cycles in S_3 (along with the identity) can be taken as a transversal for S_4/D_4.
(3) H = the *Klein four-group*; $K_4 := \{(1), (1,2)(3,4), (1,3)(2,4), (1,4)(2,3)\}$.
(4) H = R_4, the rotations of a square; $R_4 := \{(1), (1,2,3,4), (1,3)(2,4), (1,4,3,2)\}$. Note that the elements of S_3 can be taken as a transversal.

Example 9.22 Let $G = (\mathbb{R}, +)$ and let $H = (\mathbb{Z}, +)$. A transversal for H in G is in one-to-one correspondence with the interval $[0, 1)$.

9.4   Coset Representations of a Group

A construction similar to the left regular representation of a finite group G on its group algebra $\mathbb{C}[G]$ is given by the *left coset representation*.

Let H be a subgroup of a finite group G and consider G/H, the set of left cosets of H. We construct a vector space $\mathbb{C}[G/H]$ by declaring each coset $\mathbf{g_i}H$ to be a basis vector (hence the use of boldface). Thus any vector $\mathbf{v} \in \mathbb{C}[G/H]$ is of the form

$$\mathbf{v} = r_1\mathbf{g_1}H + r_2\mathbf{g_2}H + \cdots + r_k\mathbf{g_k}H, \quad \text{for scalars } r_1, \ldots, r_k.$$

The G-action is defined on the basis vectors by $g.\mathbf{g_i}H = \mathbf{gg_i}H$, and we extend this action to $\mathbb{C}[G/H]$ by linearity:

$$g.\mathbf{v} = g.(r_1\mathbf{g_1}H + r_2\mathbf{g_2}H + \cdots + r_k\mathbf{g_k}H) = r_1\mathbf{gg_1}H + r_2\mathbf{gg_2}H + \cdots + r_k\mathbf{gg_k}H.$$

As noted in Remark 9.11, only when H is a normal subgroup of G can we define multiplication in G/H, and thereby give $\mathbb{C}[G/H]$ the structure of an algebra.

Exercise* 9.23 Work through the details of the coset representations for the various subgroups of $G = S_3$: If $H = S_3$, then the coset representation can only be the trivial representation. If $H = \{(1)\}$, then the coset representation is the left regular representation on the group algebra $\mathbb{C}[G]$. When $H = S_2$ (any of the isomorphic copies), then $\mathbb{C}[G/H]$ is three dimensional and is equivalent to the permutation representation, which decomposes as $\mathcal{I} \oplus \mathcal{W}$. When $H = A_3$, the subgroup of even permutations, then $\mathbb{C}[G/H]$ is two dimensional and decomposes as $\mathcal{I} \oplus \mathcal{A}$, the direct sum of the identity and alternating representations.

Exercise 9.24 Let $G = S_4$ and $H = S_3$. Show that the coset representation of S_4 on $\mathbb{C}[G/H]$ is equivalent to the permutation representation of S_4 on $\mathcal{P}_1(x, y, z, w)$.

Exercise* 9.25 Generalize the previous exercise. Show that for any finite group G with subgroup H, the representation of G on G/H is a permutation action. Hint: If $\mathbf{g'H}$ is a basis vector in G/H, then what is $g.\mathbf{g'H}$?

9.5 Induced Representations: Version One

If H is a subgroup of G, and if W is a representation of H, what is the "most general" (whatever that means) way to extend W to a representation of G that contains a copy of W upon restriction to H? Our goal in this section is to give the "best" (to be defined later) answer.

Example 9.26 Consider the example of $H = S_2$ and $G = S_3$ acting on $\mathbb{C}[G/H]$. We can choose a transversal of H, say $G/H = \{H, (13)H, (23)H\}$, so that

$$\mathbb{C}[G/H] = \langle\!\langle \mathbf{H}, (\mathbf{13})\mathbf{H}, (\mathbf{23})\mathbf{H} \rangle\!\rangle.$$

Note that H acts trivially on $\langle\!\langle \mathbf{H} \rangle\!\rangle$, the one dimensional subspace spanned by $\{\mathbf{H}\}$, and that $\mathbb{C}[G/H] = \oplus \langle\!\langle \sigma.\mathbf{H} \rangle\!\rangle$, where the sum is taken over all distinct cosets $\sigma H \in G/H$. This is an example of a *representation of G induced from a* (trivial, in this case) *representation of H*. This generalizes.

Definition 9.27 Suppose that V is a representation of a finite group G with subgroup H, and let W be a representation of H contained in V. Then we say that the *representation of G on V is induced from the representation of H on W* if

$$V = \bigoplus_{\sigma H \in G/H} \sigma.W.$$

Again, the direct sum is taken over a transversal of H in G, and the space

$$\sigma.W := \{\sigma.w \mid w \in W\}$$

is called the *σ-translate of W*. Since the action of each $\sigma \in G$ on W is linear and invertible, each $\sigma.W$ is a subspace of V that is isomorphic (as a vector space) to W. The action of G is extended linearly over the direct sum;

$$g.\mathbf{v} = g.(\sigma_1.\mathbf{w}_1 + \cdots + \sigma_k.\mathbf{w}_k) := (g\sigma_1).\mathbf{w}_1 + \cdots + (g\sigma_k).\mathbf{w}_k.$$

Since the expression for $\mathbf{v}$ as a sum is unique, the group product is associative, and $g.\sigma_i = \sigma_j h$ for some σ_j and some $h \in H$, the action of G on $\mathbf{v}$ is well-defined.

Also, let's not overlook the fact that, since we have a coset acting on W, we need to verify that the designation $\sigma.W$ is independent of the choice of coset representative.

Exercise 9.28 Let G, H, and W be as in Definition 9.27. Work through the details of the following argument: if $aH = bH$, then $a.W = b.W$.

Let $\mathbf{w} \in W$, and consider $a.\mathbf{w}$. If $aH = bH$, then by Proposition 9.14 we have $b^{-1}a \in H$. Since W is H-invariant, we have $b^{-1}a.\mathbf{w} = \mathbf{w}'$ for some $\mathbf{w}' \in W$. Thus $a.\mathbf{w} = b.\mathbf{w}'$, so $a.W \subseteq b.W$. Similarly, $b.W \subseteq a.W$.

Notation 9.29 With the terminology of Definition 9.27, we write $V = \mathrm{Ind}_H^G W$, or $V = \mathrm{Ind}\, W$ when H and G are understood. If we need to specify the group actions, we write $\pi = \mathrm{Ind}_H^G \rho$ for ρ a representation of H and π a representation of G. Similarly we will use the notation $\chi_\pi = \chi_{\mathrm{Ind}_H^G \rho}$ or $\chi_V = \chi_{\mathrm{Ind} W}$ for *induced characters*, that is, characters of induced representations. The notation $V = W \uparrow_H^G$ or $\pi = \rho \uparrow_H^G$ is also common.

Exercise* 9.30 If H is a subgroup of a finite group G, show that the dimension of $\mathrm{Ind}_H^G W = [G : H]\,\mathrm{Dim}\, W = \frac{|G|}{|H|}\,\mathrm{Dim}\, W$.

Exercise* 9.31 Consider $\mathcal{P}_1(x, y, z) = \langle\!\langle \mathbf{x}, \mathbf{y}, \mathbf{z} \rangle\!\rangle$. Show that the permutation representation of S_3 on $\mathcal{P}_1$ is induced from a trivial representation of S_2. Try this for the three different copies of S_2 in S_3. Conclude that this induced representation is equivalent to the representation given in Example 9.26.

Exercise* 9.32 Consider $\mathcal{P}_{(1,1,0)}(x, y, z) = \langle\!\langle \mathbf{xy}, \mathbf{xz}, \mathbf{yz} \rangle\!\rangle$. Show that the permutation representation of S_3 on $\mathcal{P}_{(1,1,0)}$ is induced from a trivial representation of S_2. Try this for the three different copies of S_2 in S_3. Conclude that this induced representation is also equivalent to the representation given in Example 9.26, and consequently equivalent to the representation in Exercise 9.31.

Note that representations induced from irreducible representations need not be irreducible.

Exercise* 9.33 The vector $\mathbf{v} = \mathbf{x} - \mathbf{y}$ spans the one-dimensional alternating representation $\mathcal{A}_2$ of S_2 in $\mathcal{P}_1(x, y, z)$. Show that the space spanned by $\{\sigma.\mathbf{v} \mid \sigma \in S_3/S_2\}$ is the standard representation $\mathcal{W}_3$. Why is this NOT the induced representation $\mathrm{Ind}_{S_2}^{S_3} \mathcal{A}_2$?

Exercise 9.34 Consider the S_3-invariant subspace $\mathcal{P}_{(2,1,0)} = \langle\!\langle \mathbf{x^2y}, \mathbf{xy^2}, \ldots, \mathbf{yz^2} \rangle\!\rangle$. Certainly the vector $\mathbf{v}_1 = \mathbf{x^2y} - \mathbf{xy^2}$ spans a copy of $\mathcal{A}_2$, the alternating representation of S_2. Set $\mathbf{v}_2 = (1, 3).\mathbf{v}_1$, and $\mathbf{v}_3 = (2, 3).\mathbf{v}_1$. Check that, since

$\{(1), (13), (23)\}$ is a transversal for S_2/S_3 and $\{\mathbf{v}_1, \mathbf{v}_2, \mathbf{v}_3\}$ is linearly independent, we have $\langle\!\langle \mathbf{v}_1, \mathbf{v}_2, \mathbf{v}_3 \rangle\!\rangle = \mathrm{Ind}_{S_2}^{S_3} \mathcal{A}_2$.

Exercise* 9.35 Referring to Exercise 9.34, verify that the vector $\mathbf{v}_1 - \mathbf{v}_2 - \mathbf{v}_3$ changes sign under the action of the transpositions in S_3, and therefore spans a copy of $\mathcal{A}_3$, the alternating representation of S_3 (see Sect. 5.3) in $\mathrm{Ind}_{S_2}^{S_3} \mathcal{A}_2$. Verify that no linear combination of $\{\mathbf{v}_1, \mathbf{v}_2, \mathbf{v}_3\}$ is S_3-invariant, and therefore that $\mathrm{Ind}\,\mathcal{A}_2$ contains no copies of $\mathcal{I}_3$. It follows that $\mathrm{Ind}\,\mathcal{A}_2$ contains either two more copies of $\mathcal{A}_3$, or one copy of $\mathcal{W}_3$. Which is it?

Remark 9.36 The ways in which irreducible representations decompose into irreducibles when restricted or induced are called *branching rules*.

Example 9.37 Let G be a finite group, let $\mathbb{C}[G]$ be the group algebra, and let $H = \{e\}$ be the trivial subgroup of G. Choose any basis element $\mathbf{g} \in \mathbb{C}[G]$, and let $W = \langle\!\langle \mathbf{g} \rangle\!\rangle$ be the one-dimensional trivial representation of H. Then $\mathrm{Ind}_H^G W$ is equivalent to the left regular representation of G on $\mathbb{C}[G]$.

Example 9.38 This generalizes Example 9.37. Let H be a subgroup of a finite group G, and consider the left regular representation of H on $\mathbb{C}[H]$. Since $\mathbb{C}[H]$ has basis $\{\mathbf{h} \mid h \in H\}$, we have that $\mathrm{Ind}_H^G \mathbb{C}[H]$ is equivalent to the left regular representation of G on $\mathbb{C}[G]$.

Exercise* 9.39 Verify Example 9.38.

Exercise 9.40 Consider the S_3-invariant subspace $\mathcal{P}_{(2,1,0)} = \langle\!\langle \mathbf{x}^2\mathbf{y}, \mathbf{x}\mathbf{y}^2, \ldots, \mathbf{y}\mathbf{z}^2 \rangle\!\rangle$. Check that only the identity element in S_3 acts trivially on each basis vector, and conclude that the action of S_3 on $\mathcal{P}_{(2,1,0)}$ is induced from the (trivial) representation of the trivial subgroup $\{(1)\}$, and is therefore equivalent to the left regular representation on the group algebra.

Example 9.41 If H is a subgroup of G, if $\{W_i\}$ is a a collection of H-invariant vector spaces, and if $W = \bigoplus W_i$, then $\mathrm{Ind}\,W = \bigoplus \mathrm{Ind}\,W_i$. This is routine to verify using the definitions involved.

Example 9.42 From the previous example, it follows that if W is a representation of H and if W_1 is an H-invariant subspace of W, then $\mathrm{Ind}_H^G W_1$ is a G-invariant subspace of $\mathrm{Ind}_H^G W$.

Exercise* 9.43 In this exercise we will decompose representations induced from trivial representations of a subgroup H of S_4 (for the listed subgroups H) by considering the representation of S_4 on the coset space $\mathbb{C}[S_4/H]$, applying Exercises 7.6 and 9.25, and using characters.

For each of the subgroups H below:

- Determine the dimension of $\mathbb{C}[S_4/H]$.
- Write out the cosets of S_4/H, or at least determine a transversal (Definition 9.15). Some coset representatives are easier to work with than others.
- Observe that the action of S_4 is a permutation action on the cosets. Using Exercise 7.19, compute the character of the representation of S_4 on $\mathbb{C}[S_4/H]$.
- Using Exercise 7.19 again, determine the multiplicity of each irreducible representation of S_4 in $\mathbb{C}[S_4/H]$. Hint: The characters for the trivial and alternating representations are easy since their eigenvalues are obvious. As you work through the list of subgroups, new irreducible characters of S_4 will arise.

(1) $H = A_4$, the alternating subgroup of S_4.
(2) $H = D_4$, the dihedral subgroup from Exercise 1.53.
(3) $H = S_3$.
(4) $H = K_4$, the *Klein four-group*, $K_4 := \{(1), (1, 2)(3, 4), (1, 3)(2, 4), (1, 4)(2, 3)\}$.
(5) $H = R_4$, the rotations of a square, $R_4 := \{(1), (1, 2, 3, 4), (1, 3)(2, 4), (1, 4, 3, 2)\}$.
(6) $H = A_3$, the alternating subgroup of S_3.

9.6 Matrix Realizations and Characters of Induced Representations

If H is a subgroup of a finite group G, and if W is a representation of H, we can obtain matrix realizations and characters of Ind W from those of W.

Let $S = \{\sigma_i\}_{i=1}^k$ be a transversal for G/H. Then any $\mathbf{v} \in V$ can be written as $\mathbf{v} = \sum \sigma_i.\mathbf{w}_i$ for $\mathbf{w}_i \in W$, and therefore $g.\mathbf{v} = \sum g\sigma_i.\mathbf{w}_i$. How does $g \in G$ act on some $\sigma_i.\mathbf{w}_i$? Since G/H partitions G, we have that $g.\sigma_i = \sigma_j h$ for a unique $\sigma_j \in S$ and some $h \in H$, which tells us how $g.\sigma_i = \sigma_j h$ acts: first we get an action of H on W, then a translation by σ_j from W to $\sigma_j W$. Now $g.\sigma_i = \sigma_j h$ implies that $\sigma_j^{-1} g.\sigma_i = h$. Thus the matrix realization for $g \in G$ is given by block matrices, where in each ith column there is a unique j with block $\overbrace{\sigma_j^{-1} g.\sigma_i}$, the other blocks in the column being zero since in those cases $\sigma_{j'}^{-1} g.\sigma_i \notin H$. This yields the following proposition and corollary.

Proposition 9.44 *Let H be a subgroup of a finite group G, and suppose W is a d-dimensional representation of H. Let $S = \{\sigma_i\}_{i=1}^k$ be a transversal for G/H, and let $V = \mathrm{Ind}_H^G W$. Then the matrix realization $\tilde{g}$ for the action of g on V is given in* *block form by*

$$
\tilde{g} = \begin{pmatrix}
\widetilde{\sigma_1^{-1} g \sigma_1} & \widetilde{\sigma_1^{-1} g \sigma_2} & \cdots & \widetilde{\sigma_1^{-1} g \sigma_k} \\
\widetilde{\sigma_2^{-1} g \sigma_1} & \widetilde{\sigma_2^{-1} g \sigma_2} & \cdots & \widetilde{\sigma_2^{-1} g \sigma_k} \\
\vdots & \cdots & \ddots & \vdots \\
\widetilde{\sigma_k^{-1} g \sigma_1} & \widetilde{\sigma_k^{-1} g \sigma_2} & \cdots & \widetilde{\sigma_k^{-1} g \sigma_k}
\end{pmatrix},
$$

where each block is a $d \times d$ matrix, and exactly one block in each row (or column) is non-zero.

The results of Proposition 9.44 simplify the task of computing characters of induced representations.

Corollary 9.45

$$
\chi_V(g) = \sum_{i=1}^{k} \mathrm{tr}\left(\widetilde{\sigma_i^{-1} g \sigma_i} \right) = \sum_{i=1}^{k} \chi_W(\sigma_i^{-1} g \sigma_i), \quad \text{where } \chi_W(\sigma_i^{-1} g \sigma_i) := 0 \text{ whenever } \sigma_i^{-1} g \sigma_i \notin H.
$$

Exercise* 9.46 Work out some matrix realizations for $\mathrm{Ind}_{S_2}^{S_3} \mathcal{I}_2$ and $\mathrm{Ind}_{S_2}^{S_3} \mathcal{A}_2$. Verify the character for $\mathrm{Ind}_{S_2}^{S_3} \mathcal{A}_2$ from Exercise 9.35.

Exercise 9.47 Show that, since characters are constant on conjugacy classes, we can rewrite Corollary 9.45 as

$$
\chi_V(g) = \frac{1}{|H|} \sum_{x \in G} \chi_W(x^{-1} g x), \quad \text{where again } \chi_W(x^{-1} g x) := 0 \text{ whenever } x^{-1} g x \notin H.
$$

9.7 Construction of Induced Representations

Definition 9.27 tells us <u>when</u> a representation of a finite group G is induced from a representation of one of its subgroups H. By reverse-engineering the discussion from the previous section, we can construct the representation of G induced from the representation of H.

Suppose W be a representation of H, and let $S = \{\sigma_i\}_{i=1}^{k}$ be a transversal for G/H. For each $\sigma_i \in S$, take a copy W_{σ_i} of W, and then set $V = \bigoplus_{i=1}^{k} W_{\sigma_i}$ (this would be an *external* direct sum). How does G act on V? For any $g \in G$, we have $g = \sigma_j h$ for some $\sigma_j \in S$ and some $h \in H$. Then for any $\mathbf{w}_i \in W_{\sigma_i}$, we have $g.\mathbf{w}_i = \sigma_j h.\mathbf{w}_i$, so g first acts on $\mathbf{w}_i$ via h, then shifts the result to the direct summand $W_{\sigma_j \sigma_i}$. Now extend by linearity to all of V. Of course, this will give us a matrix realization equivalent to the one in Proposition 9.44.

9.8   Frobenius Reciprocity

In this section we relate restricted and induced representations.

Proposition 9.48 *Suppose G is a finite group with subgroup H. Suppose U is a representation of G, W a representation of H, and let $V = \operatorname{Ind} W$. Then every H-linear map $\varphi \colon W \to U$ extends uniquely to a G-linear map $\widehat{\varphi} \colon V \to U$. Conversely, every G-linear map from V to U restricts to a unique H-linear map from W to $\operatorname{Res} U$. Consequently, the identification of φ with $\widehat{\varphi}$ gives an isomorphism of vector spaces;*

$$\operatorname{Hom}_H(W, \operatorname{Res} U) \cong \operatorname{Hom}_G(\operatorname{Ind} W, U).$$

Proof Let $\varphi \colon W \to U$ be H-linear, and let $\{\sigma_i\}_{i=1}^k$ be a transversal for G/H. Since $V = \operatorname{Ind} W = \bigoplus_{i=1}^k \sigma_i.W$, every $\mathbf{v} \in V$ can be written uniquely as $\mathbf{v} = \sigma_1.\mathbf{w}_1 + \cdots + \sigma_k.\mathbf{w}_k$ for some vectors $\mathbf{w}_i \in W$. Define $\widehat{\varphi} \colon V \to U$ by

$$\widehat{\varphi}(\mathbf{v}) := \sigma_1.\varphi(\mathbf{w}_1) + \cdots + \sigma_k.\varphi(\mathbf{w}_k).$$

Because the sum is direct and every linear map from W to U is uniquely determined by its values on the basis vectors of W, it follows that $\widehat{\varphi}$ is the unique extension of φ from W to $V = \operatorname{Ind} W$.

The unique restriction of a G-linear map to an H-linear map should be obvious. $\qquad\square$

Exercise* 9.49 Verify that the map $\widehat{\varphi}$ in Proposition 9.48 is well-defined.

Exercise 9.50 Verify that the map $\widehat{\varphi}$ in Proposition 9.48 is G-linear.

Proposition 9.48 has a useful corollary.

Corollary 9.51 *If W and U are irreducible, then the multiplicity of U in $\operatorname{Ind} W$ is equal to the multiplicity of W in $\operatorname{Res} U$.*

Proof If W and U are irreducible, then   by Proposition 5.16, the dimension of $\operatorname{Hom}_G(\operatorname{Ind} W, U)$ is the number of times U appears in $\operatorname{Ind} W$, while the dimension of $\operatorname{Hom}_H(W, \operatorname{Res} U)$ is the number of times W appears in $\operatorname{Res} U$.

This result is often stated and proven in terms of inner products of characters. See [Sa], Theorem 1.12.6. $\qquad\square$

Remark 9.52 For the more advanced reader, a stronger statement of Frobenius Reciprocity says that the operations of restriction and induction are adjoint functors between the category of G-modules and the category of H-modules. In this context, Proposition 9.48 is a "universal property" that uniquely determines $\operatorname{Ind} W$ up to a natural G-isomorphism. This is what we meant by the term "best" at the beginning of Sect. 9.5.

Just because we define something does not necessarily mean that it exists. We can, however, guarantee the existence and uniqueness of induced representations.

Proposition 9.53 *Given a subgroup H of a group G and a representation W of H, the representation $\mathrm{Ind}_H^G W$ exists and is unique up to equivalence.*

Proof We first show existence. By Example 9.41, we may assume that W is irreducible. By Corollary 7.19 (5) or Remark 6.4, W is a direct summand of $\mathbb{C}[H]$. By Examples 9.38 and 9.42, Ind W is a G-invariant subspace of $\mathbb{C}[G] = \mathrm{Ind}\,\mathbb{C}[H]$.

We now show uniqueness, up to equivalence. Let H be a subgroup of G, let W be a representation of H, and suppose V and V' are two representations of G induced from the representation W of H. Let $i \colon W \to V'$ be the inclusion mapping[2] of W into V' as a direct summand. Since i is obviously H-linear, by Proposition 9.48 there is a unique G-linear map $F \colon V \to V'$ that intertwines the representations of G on V and V', and is the identity on W. Therefore the image of F contains each $\sigma.W$, and consequently is equal to all of V'. Since V and V' have the same dimension (Exercise 9.30), F is an isomorphism of vector spaces. Since F is also a G-map, we have that V and V' are equivalent as representations of G. $\square$

9.9 Induced Representations: Version Two

In this section we give another definition of induced representations, and show that, at least in the case for finite groups, these definitions are equivalent.

Definition 9.54 Let G be a group with subgroup H, let W be a representation of H, and define

$$\mathcal{F} := \{f \colon G \to W \mid h^{-1}.[f(x)] = f(xh) \text{ for all } h \in H \text{ and } x \in G\}.$$

The action of G on $\mathcal{F}$ is then defined by $[g.f](x) := f(g^{-1}x)$. It is important to note that the expression $h^{-1}.[f(x)]$ refers to the action of H on W, while the expression $[g.f](x)$ refers to the action of G on $\mathcal{F}$.

Exercise 9.55 If $f \in \mathcal{F}$, show that $g.f \in \mathcal{F}$. Conclude, along with Exercise 1.92, that the space $\mathcal{F}$ carries a representation of G.

Definition 9.54 has the advantage that it can be applied to groups that, say, are not finite, or to subgroups that are not of finite index; for example, matrix groups and

---

[2] If $X \subset Y$, then the inclusion mapping $i \colon X \hookrightarrow Y$ is defined as $i(x) = x$ for all $x \in X$.

subgroups. However, this definition if not very intuitive. What must such a function $f \in \mathcal{F}$ "look like?"

Since $f: G \to W$, we have that $f(e) = \mathbf{w}$ for some $\mathbf{w} \in W$. Applying the definition of $\mathcal{F}$ yields

$$h^{-1}.[f(e)] = f(eh) = f(h) = h^{-1}.\mathbf{w}, \quad \text{where} \quad \mathbf{w} = f(e).$$

In other words, for any such $f \in \mathcal{F}$, the image $f(h)$ is a "left action by h on some $\mathbf{w} \in W$," at least when restricted to H. We can then extend this idea to all of G via cosets of H.

The following exercises flesh out the details and outline the proof that the above representation of a finite group G on $\mathcal{F}$ is equivalent to the induced representation of G on $V = \mathrm{Ind}_H^G W = \bigoplus_{\sigma H \in G/H} \sigma.W$.

Exercise* 9.56 Let H be a subgroup of a finite group G, and let W be a representation of H. For $\mathbf{w} \in W$ and $x \in G$, define $f_{\mathbf{w}}: G \to W$ by

$$f_{\mathbf{w}}(x) := \begin{cases} x^{-1}.\mathbf{w}, & \text{if } x \in H; \\ 0, & \text{if } x \notin H. \end{cases}$$

Show that $f_{\mathbf{w}} \in \mathcal{F}$.

Exercise* 9.57 Define $\Psi: W \to \mathcal{F}$ by $\Psi(\mathbf{w}) = f_{\mathbf{w}}$. Show that Ψ is an H-linear map.

Exercise* 9.58 Show that the map $\Psi: W \to \mathcal{F}$ is one-to-one. Conclude that Ψ is an H-map that embeds W in $\mathcal{F}$ as $\Psi(W) = \widehat{W} = \{f_{\mathbf{w}} \mid \mathbf{w} \in W\}$, and therefore W and $\widehat{W}$ are equivalent representations of H.

Remark 9.59 In mathematics, an *embedding* is when one mathematical structure is contained within another such structure: subgroups of a group, vector subspaces of a vector space, and topological subspaces of a topological space are some common examples. When some object X is said to be embedded in another such object Y, the embedding is given by some injective (one-to-one) and structure-preserving map $f: X \to Y$. Respectively, group homomorphisms, linear maps between vector spaces, and homeomorphisms of topological spaces are the structure-preserving maps for the common examples named above.

Exercise* 9.60 We now extend the notion of $f_{\mathbf{w}}$ to the other cosets of H. Choose $g \in G$, $\mathbf{w} \in W$, and define $f_{g,\mathbf{w}}: G \to W$ as

$$f_{g.\mathbf{w}}(x) := \begin{cases} h^{-1}.\mathbf{w}, & \text{if } x = gh \text{ for some } h \in H; \\ 0, & \text{if } x \notin gH. \end{cases}$$

Show that $f_{g.\mathbf{w}} \in \mathcal{F}$.

Exercise* 9.61 Let $\sigma \in G$. Show that $\sigma.f_{g.\mathbf{w}} = f_{\sigma g.\mathbf{w}}$. Note that $f_{\mathbf{w}} = f_{e.\mathbf{w}}$.

Exercise* 9.62 For $\sigma \in G$, define $\sigma.\widehat{W} := \{\sigma.f_{\mathbf{w}} \mid \mathbf{w} \in W\}$, and define $\widehat{V} := \bigoplus_{\sigma H \in G/H} \sigma.\widehat{W}$. Show that $\mathcal{F} = \widehat{V}$. That is, given any $f \in \mathcal{F}$, then $f = f_{\sigma.\mathbf{w}}$ for some $\sigma \in G$ and some $\mathbf{w} \in W$.

Exercises 9.56–9.62 prove the following.

Proposition 9.63 *Suppose that H is a subgroup of a finite group G, and let W be a representation of H. Let $\mathcal{F}$ be defined as in Definition 9.54, and let $f_{\mathbf{w}}$ be defined as in Exercise 9.56. Define $\widehat{W} := \{f_{\mathbf{w}} \mid \mathbf{w} \in W\}$, and define $\sigma.\widehat{W} := \{\sigma.f_{\mathbf{w}} \mid \mathbf{w} \in W\}$. Then*

$$\mathcal{F} = \sum_{\sigma H \in G/H} \sigma.\widehat{W} \quad and \quad V = \bigoplus_{\sigma H \in G/H} \sigma.W$$

are equivalent representations of G.

The following exercises provide a streamlined proof that W and $\widehat{W}$ are equivalent as representations of H.

Exercise* 9.64 Let $\widehat{W} = \{f_{\mathbf{w}} \mid \mathbf{w} \in W\}$. Define $\Phi \colon \widehat{W} \to W$ given by $\Phi(f) = f(e)$, where e is the identity in G. Show that Φ is also an H-linear map.

Exercise* 9.65 With Φ as above and Ψ from Exercise 9.57, show that $\Phi \circ \Psi = Id_W$ and $\Psi \circ \Phi = Id_{\widehat{W}}$. Conclude that Ψ embeds W in $\mathcal{F}$ as $\widehat{W}$, and that W and $\widehat{W}$ are equivalent as representations of H.

Example 9.66 Here is a special case of Exercise 9.65.

Let $H = G$ with identity element e, and let W be a representation of G. Define $\Phi \colon \mathcal{F} \to W$ by $f \mapsto f(e)$. Then[3]

$$\Phi[g.f] = [g.f](e) = f(g^{-1}e) = f(eg^{-1}) = g.[f(e)] = g.\Phi(f).$$

That is, the map Φ intertwines the action of G on $\mathcal{F}$ with the action of G on W.

For $\mathbf{w} \in W$ and $x \in G$, define $f_{\mathbf{w}}(x) := x^{-1}.\mathbf{w}$, and define $\Psi \colon W \to \mathcal{F}$ by $\Psi(\mathbf{w}) = f_{\mathbf{w}}$. Then Φ also intertwines the action of G on W with the action of G on $\mathcal{F}$, and the maps Φ and Ψ are inverses of each other. Hence W and $\mathcal{F}$ are equivalent representations of G.

Exercise* 9.67 Prove the following statement of Frobenius reciprocity for the version of induced representations given in Definition 9.54.

---

[3] Because $f \in \mathcal{F}$ and $H = G$.

Let H be a subgroup of a group G with identity e, let (ρ, W) be a representation of H, let (π, U) be a representation of G, and let $\mathcal{F} = \text{Ind}_H^G W$. Then there is an isomorphism of vector spaces;

$$\text{Hom}_H(\text{Res } U, W) \cong \text{Hom}_G(U, \text{Ind } W).$$

Hint: Show that any $\Phi \in \text{Hom}_G(U, \text{Ind } W)$ corresponds to a map $\phi \in \text{Hom}_H(\text{Res } U, W)$; where Φ is expressed in terms of ϕ, and ϕ is expressed in terms of Φ, as

$$\phi(u) = [\Phi(u)](e), \quad \text{and} \quad [\Phi(u)](x) = \phi(\pi[x^{-1}]u), \quad \text{for } x \in G.$$

Remark 9.68 There are a number of (equivalent) definitions of induced representations. The matrix realization of Proposition 9.44, and the space $\mathcal{F}$ described above are often given as the definition of the representation $\text{Ind}_H^G W$, as is the variation of the space $\mathcal{F}$ that we'll call

$$\mathcal{F}_{op} = \{f : G \to W \mid h.[f(x)] = f(hx) \text{ for all } h \in H \text{ and } x \in G\}.$$

The action of G on $\mathcal{F}_{op}$ is given by $[g.f](x) = f(xg)$, and Proposition 9.63 is stated and proven "backwards." Another definition uses tensor products, and some references define $\text{Ind}_H^G W$ as any representation that satisfies the "unique extension property" of Proposition 9.48. For the geometric version of induced representations, on the space of sections on a vector bundle over the quotient space G/H (with structure group H and fiber W), along with some physical motivation for the idea, see [St], Ch 3.

9.10   Hints and Additional Comments

Exercise 9.5 We use the basis

$$\mathcal{W}_4 = \{r_1\mathbf{x} + r_2\mathbf{y} + r_3\mathbf{z} + r_4\mathbf{w} \mid r_1 + r_2 + r_3 + r_4 = 0\} = \langle\!\langle \mathbf{x} - \mathbf{y}, \ \mathbf{x} + \mathbf{y} - 2\mathbf{z}, \ \mathbf{x} + \mathbf{y} + \mathbf{z} - 3\mathbf{w} \rangle\!\rangle.$$

It follows that, when restricted to S_3, the representation $\text{Res } \mathcal{W}_4$ decomposes as

$$\langle\!\langle \mathbf{x} - \mathbf{y}, \mathbf{x} + \mathbf{y} - 2\mathbf{z} \rangle\!\rangle \oplus \langle\!\langle \mathbf{x} + \mathbf{y} + \mathbf{z} - 3\mathbf{w} \rangle\!\rangle \cong \mathcal{W}_3 \oplus \mathcal{I}_3.$$

Note that when we further restrict to S_2, $\text{Res } \mathcal{W}_4$ decomposes as $\mathcal{A}_2 \oplus 2\,\mathcal{I}_2$.

Exercise 9.6 Here are a few solutions that demonstrate the techniques involved.

(1) If $\mathbf{v} + W = \mathbf{v}' + W$, then for each $\mathbf{w} \in W$ there is a $\mathbf{w}' \in W$ such that $\mathbf{v} + \mathbf{w} = \mathbf{v}' + \mathbf{w}'$, so $\mathbf{v} - \mathbf{v}' = \mathbf{w}' - \mathbf{w} \in W$ since W is a subspace.

Conversely, suppose $\mathbf{v} - \mathbf{v}' \in W$. Then $\mathbf{v} - \mathbf{v}' = \mathbf{w}'$ for some $\mathbf{w}' \in W$, and so $\mathbf{v} = \mathbf{v}' + \mathbf{w}'$. Therefore, for each $\mathbf{w} \in W$, we have

$$\mathbf{v} + \mathbf{w} = \mathbf{v}' + \mathbf{w}' + \mathbf{w} \in \mathbf{v}' + W, \quad \text{so} \quad \mathbf{v} + W \subset \mathbf{v}' + W.$$

By a symmetric argument, $\mathbf{v}' + W \subset \mathbf{v} + W$.

(2) We need to show that if $\mathbf{v} + W = \mathbf{v}' + W$, and if $\mathbf{u} + W = \mathbf{u}' + W$, then

$$[\mathbf{v} + W] + [\mathbf{u} + W] = [\mathbf{v}' + W] + [\mathbf{u}' + W],$$

i.e, that addition is independent of the coset representatives chosen. The proof is a routine application of the definitions involved and part (1).

Similarly for scalar multiplication. For example;

$$
\begin{aligned}
\mathbf{v} + W = \mathbf{v}' + W &\Rightarrow \mathbf{v} - \mathbf{v}' = \mathbf{w} \in W \\
&\Rightarrow r(\mathbf{v} - \mathbf{v}') = r\mathbf{w} \in W \quad \text{(since } W \text{ is a subspace)} \\
&\Rightarrow r\mathbf{v} - r\mathbf{v}' \in W \\
&\Rightarrow r\mathbf{v} + W = r\mathbf{v}' + W \quad \text{(by part (1))} \\
&\Rightarrow r[\mathbf{v} + W] = r[\mathbf{v}' + W].
\end{aligned}
$$

(6) $\quad$
$$
\begin{aligned}
\mathbf{v} + W = \mathbf{v}' + W &\Rightarrow \mathbf{v} - \mathbf{v}' = \mathbf{w} \in W \\
&\Rightarrow g.(\mathbf{v} - \mathbf{v}') = g.\mathbf{w} \in W \text{ (since } W \text{ is } G\text{-invariant)} \\
&\Rightarrow g.\mathbf{v} - g.\mathbf{v}' \in W \\
&\Rightarrow g.\mathbf{v} + W = g.\mathbf{v}' + W \\
&\Rightarrow g.[\mathbf{v} + W] = g.[\mathbf{v}' + W].
\end{aligned}
$$

Also,

$$
\begin{aligned}
\phi(g.\mathbf{v}) &= g.\mathbf{v} + W \\
&= g.[\mathbf{v} + W] \\
&= g.\phi(\mathbf{v}).
\end{aligned}
$$

Exercise 9.7 As with all maps on equivalence classes, we first need to show that $\widehat{\phi}$ is well defined, so suppose that $\mathbf{v}_1 + K = \mathbf{v}_2 + K$. Then $\mathbf{v}_1 = \mathbf{v}_2 + \mathbf{k}$ for some $\mathbf{k} \in K$, and therefore,

$$\widehat{\phi}(\mathbf{v}_1 + K) = \phi(\mathbf{v}_1) = \phi(\mathbf{v}_2 + \mathbf{k}) = \phi(\mathbf{v}_2) + \phi(\mathbf{k}) = \phi(\mathbf{v}_2) + 0 = \widehat{\phi}(\mathbf{v}_2 + K).$$

Now recall that $\operatorname{Ker} \phi$ and $\operatorname{Im} \phi$ are G-invariant subspaces of V and W respectively (Proposition 3.13). It should be routine to verify that $\widehat{\phi}$ is linear, onto $\operatorname{Im} \phi$, and a G-map.

We can show that $\widehat{\phi}$ is one-to-one directly:

$$
\begin{aligned}
\widehat{\phi}(\mathbf{v}_1 + K) = \widehat{\phi}(\mathbf{v}_2 + K) &\Rightarrow \phi(\mathbf{v}_1) = \phi(\mathbf{v}_2) \\
&\Rightarrow \phi(\mathbf{v}_1) - \phi(\mathbf{v}_2) = 0 \\
&\Rightarrow \phi(\mathbf{v}_1 - \mathbf{v}_2) = 0 \\
&\Rightarrow \mathbf{v}_1 - \mathbf{v}_2 \in K \\
&\Rightarrow \mathbf{v}_1 + K = \mathbf{v}_2 + K.
\end{aligned}
$$

Or we can just observe that $\operatorname{Ker} \widehat{\phi} = K$, which is the zero-element in V/K. The conclusion that $V \cong \operatorname{Im} \phi \oplus K$ follows from Exercise 9.6.

Exercise 9.8 First observe that the set $\{g.\mathbf{v} \mid g \in G\}$ spans an invariant subspace of V, and so equals all of V since $\mathbf{v} \neq 0$ and V is irreducible. Now verify that ϕ is a G-map, that $\operatorname{Im} \phi = V$, and so $\mathbb{C}[G]$ contains a copy of V as a direct summand by Exercise 9.7.

Exercise 9.16

(1) Following the reasoning of Example 9.12, coset representatives are character-ized by their determinant. Hence the set $\{\left[\begin{smallmatrix} r & 0 \\ 0 & 1 \end{smallmatrix}\right] \mid r \in \mathbb{R}^*\}$ is a system of representatives for $GL(2, \mathbb{R})/SL(2, \mathbb{R})$.
(2) The set $\{0, 1, 2\}$ is a transversal for $\mathbb{Z}/3\,\mathbb{Z}$ since cosets are determined by the remainder after division by 3.

Exercise 9.17 For the novice reader, this example is important enough to justify some additional remarks. If $G = (\mathbb{Z}, +)$ and $H = (k\,\mathbb{Z}, +)$, then G/H is known as the *integers modulo k*, often written as $\mathbb{Z}_k$. For example $\mathbb{Z}_6 = \{6n, 1 + 6n, \ldots, 5 + 6n \mid n \in \mathbb{Z}\}$ (recall that here the cosets are written additively), which we can abbreviate as $\{\bar{0}, \bar{1}, \ldots, \bar{5}\}$. The set $\mathbb{Z}_k$ has the algebraic structure of a *ring*, which means that we can add, subtract and multiply (among other things). For example, in $\mathbb{Z}_6$ we have $\bar{4} + \bar{3} = \bar{1}$ and $\bar{4} \times \bar{2} = \bar{2}$. It is a good exercise to show that these operations are well defined, *i.e.*, are independent of the coset representatives chosen. Note that $\bar{4} + \bar{2} = \bar{0}$, which means that $\bar{2} = -\bar{4}$. Also observe that $\bar{2} \times \bar{3} = \bar{0}$; the cosets $\bar{2}$ and $\bar{3}$ are called *zero divisors*. Furthermore $\bar{5} \times \bar{5} = \bar{1}$ so $\bar{5}^{-1} = \bar{5}$, but that there is no $\bar{2}^{-1}$, so we can't always divide. It turns out that if p is prime, then $\mathbb{Z}_p$ has the algebraic structure of a *field*. One common and useful way to think of, say, $\mathbb{Z}_6$, is as a clock face numbered $1, \ldots, 6 = 0$, rather than $1, \ldots, 12$.

Exercise 9.18 For $G = (\mathbb{C}^*, \times)$ and $H = \{a + bi \mid a^2 + b^2 = 1\}$, we will describe the coset $(1 + 2i)H$. By Proposition 9.14, $gH = g'H$ if and only if $g^{-1}g' \in H$, so we need to determine all g' such that $(1 + 2i)^{-1}g' \in H$. Now $(1 + 2i)^{-1} = \frac{1}{5} - \frac{2}{5}i$, and if we set $g' = x + yi$, then

$$g^{-1}g' = (\tfrac{1}{5} - \tfrac{2}{5}i)(x + yi)$$
$$= (\tfrac{1}{5}x + \tfrac{2}{5}y) + (\tfrac{-2}{5}x + \tfrac{1}{5}y)i.$$

To satisfy the condition $a^2 + b^2 = 1$ for membership in H, we solve

$$(\tfrac{1}{5}x + \tfrac{2}{5}y)^2 + (\tfrac{-2}{5}x + \tfrac{1}{5}y)^2 = 1,$$

and obtain $x^2 + y^2 = 5$. Thus $(1 + 2i)H = \{z \in \mathbb{C} \mid |z| = \sqrt{5}\}$.

For those readers familiar with complex variables it is easier to solve this more generally. Write $g \in \mathbb{C}$ in polar form as $g = r(\cos\theta + i\sin\theta)$, where r is the *modulus*: $r = (g\overline{g})^{1/2}$ (the distance from the origin to g in the complex plane), and where θ is the angle between the positive real axis and the ray from the origin through g. We can similarly write $g' = r'(\cos\theta' + i\sin\theta')$. It follows that

$$g^{-1}g' = [r(\cos\theta + i\sin\theta)]^{-1}\, r'(\cos\theta' + i\sin\theta')$$
$$= \tfrac{1}{r}(\cos\theta - i\sin\theta)\, r'(\cos\theta' + i\sin\theta')$$
$$= \tfrac{r'}{r}(\cos[\theta' - \theta] + i\sin[\theta' - \theta]) \in H.$$

Since $\cos^2\phi + \sin^2\phi = 1$ for any real number ϕ, the right-most factor lies in H. Therefore g and g' are in the same coset of H if and only if $r = r'$, *i.e.*, if and only if they have the same modulus. It follows that G/H partitions G, *i.e.*, the Complex plane $/\{0\}$, into circles centered at 0.

Exercise 9.19 For $G = (\mathbb{R}^2, +)$, $H = \{(x, 2x) \mid x \in \mathbb{R}\}$ and $g = (1, 3) \in G$, we wish to find all $g' = (a, b)$ such that $g + H = g' + H$, or equivalently $(a, b) - (1, 3) \in H$. To wit,

$$(a, b) - (1, 3) \in H \Leftrightarrow (a, b) - (1, 3) \in \{(x, 2x) \mid x \in \mathbb{R}\}$$
$$\Leftrightarrow (a - 1, b - 3) \in \{(x, 2x) \mid x \in \mathbb{R}\}$$
$$\Leftrightarrow b - 3 = 2(a - 1)$$
$$\Leftrightarrow b = 2a + 1.$$

Thus $(1, 3) + H = \{(a, 2a+1) \mid a \in \mathbb{R}\}$, or more conventionally, the line $y = 2x+1$ in the xy-plane. Similar reasoning shows that G/H partitions $\mathbb{R}^2$ into parallel lines with slope 2.

Exercise 9.23 The case where $H = A_3$ may not be obvious. $S_3/A_3 = \{A_3, (1, 2).A_3\}$, so the identity representation is spanned by $\{A_3 + (1, 2).A_3\}$, and the alternating representation is spanned by $\{A_3 - (1, 2).A_3\}$.

Exercise 9.25 The relevant observation is that, for any $g \in G$ and any $g'H \in G/H$, the G-action given by $g.g'H$ is another coset in G/H. That is, the action of G just permutes the cosets, which are the basis vectors in $\mathbb{C}[G]$.

Exercise 9.30 By the definition of direct sum, the spaces $\sigma.W$ intersect trivially, and each σ-translate of W has the same dimension. By Lagrange's theorem, $[G : H] = \frac{|G|}{|H|}$.

Exercises 9.31 and 9.32 The standard embedding of S_2 in S_3 acts trivially on the subspace $\ll \mathbf{z} \gg$ of $\mathcal{P}_1$, and trivially on the subspace $\ll \mathbf{xy} \gg$ of $\mathcal{P}_{(1,1,0)}$.

Exercise 9.33 The set $\{\sigma.(\mathbf{x} - \mathbf{y}) \mid \sigma \in S_3/S_2\}$ is not linearly independent, so you will need to reduce the span of this set to obtain a basis for $\mathcal{W}$.

This is not the induced representation because the sum is not the **direct** sum of the one-dimensional spaces $\sigma.(\mathbf{x} - \mathbf{y})$.

Exercise 9.35 Following Sect. 8.5, we use the Young symmetrizer corresponding to

$$T = \begin{array}{|c|c|} \hline 1 & 2 \\ \hline 3 \\ \cline{1-1} \end{array}$$

to obtain $\mathbf{E}_T.\mathbf{v}_1 = -\mathbf{v}_2 + \mathbf{v}_3 := \mathbf{w}_1$. Now $(1,2,3).\mathbf{w}_1 = -\mathbf{v}_1 - \mathbf{v}_3 := \mathbf{w}_2$ which is not in the span of $\mathbf{w}_1$, so $\{\mathbf{w}_1, \mathbf{w}_2\}$ spans a copy of $\mathcal{W}_3$ in $\mathrm{Ind}\,\mathcal{A}_2$. It follows that $\mathrm{Ind}\,\mathcal{A}_2 \simeq \mathcal{A}_3 \oplus \mathcal{W}_3$.

Exercise 9.39 Let $H = \{h_1, \ldots, h_k\}$, and let $\{g_1, \ldots, g_\ell\}$ be a cross-section of G/H. Our goal is to show that

$$\mathbb{C}[G] = \bigoplus_{i=1}^{\ell} g_i \mathbb{C}[H].$$

Since G is the disjoint union of its H-cosets, it follows that $G = \{g_i h_j\}$ for $i = 1 \ldots \ell$, and $j = 1 \ldots k$. Hence the right-hand side spans $\mathbb{C}[G]$.

To show that the sum is direct, note that the set $\{\mathbf{gh}_1, \ldots, \mathbf{gh}_k\}$ is a basis for $g.\mathbb{C}[H]$, the set $\{\mathbf{g'h}_1, \ldots, \mathbf{g'h}_k\}$ is a basis for $g'.\mathbb{C}[H]$, and these two sets are disjoint whenever $gH \neq g'H$, in which case $g.\mathbb{C}[H] \bigcap g'.\mathbb{C}[H] = \{0\}$. We could also just add up the dimensions.

Exercise 9.43 To decompose these coset representations using characters, we make a character table as in Sect. 7.2. Across the top is a representative from each conjugacy class, and the number in brackets is the number of elements in each such class.

$\mathbb{C}[\mathbf{S_4}/\mathbf{A_4}]$. We have $S_4/A_4 = \{A_4, (1,2).A_4\}$, and since the action of S_4 permutes these cosets, the character of this representation is the number of cosets fixed by each conjugacy class. Since the even permutations fix both cosets, and the odd permutations fix neither, we have the following character table.

	[1]	[6]	[8]	[6]	[3]
S_4	(1)	(1, 2)	(1, 2, 3)	(1, 2, 3, 4)	(1, 2)(3, 4)
$\mathbb{C}[S_4/A_4]$	2	0	2	0	2

It should be clear that there is at least one copy of the trivial representation (spanned by the orbit sum), and using property (1) from Proposition 7.4 allows us to compute the character of the one-dimensional complimentary subspace, and whose character reveals it to be the alternating representation.

	[1]	[6]	[8]	[6]	[3]
S_4	(1)	(1, 2)	(1, 2, 3)	(1, 2, 3, 4)	(1, 2)(3, 4)
$\mathbb{C}[S_4/A_4]$	2	0	2	0	2
$\mathcal{I}_4 \cong V^{(4)}$	1	1	1	1	1
$\mathcal{A}_4 \cong V^{(1,1,1,1)}$	1	-1	1	-1	1

We can write out these subspaces explicitly; $\{\mathbf{A_4} + (\mathbf{1,2}).\mathbf{A_4}\}$ spans the trivial representation, and $\{\mathbf{A_4} - (\mathbf{1,2}).\mathbf{A_4}\}$ spans the alternating representation.

$\mathbb{C}[\mathbf{S_4}/\mathbf{D_4}]$. There are lots of ways to designate the cosets of S_4/D_4. One simple way is to use 2-cycles; $S_4/D_4 = \{D_4, (1,2).D_4, (2,3).D_4\}$. To determine which cosets are fixed by S_4, you can either write out the cosets, or use property (5) from Proposition 9.14 and the fact that conjugation preserves cycle classes. For example,

$$(1, 2, 3).[(1, 2).D_4] \neq (1, 2).D_4, \quad \text{because} \quad (1, 2)^{-1}[(1, 2, 3)(1, 2)] = (1, 3, 2) \notin D_4$$

since D_4 contains no 3-cycles. Subtracting off the trivial character yields the character for the complimentary 2-dimensional subspace, which is irreducible by Exercise 7.19, part (3). If you jump ahead to Corollary 11.21 you'll see that there is only one two-dimensional irreducible representation of S_4, labeled by the partition $(2, 2)$.

	[1]	[6]	[8]	[6]	[3]
S_4	(1)	(1, 2)	(1, 2, 3)	(1, 2, 3, 4)	(1, 2)(3, 4)
$\mathbb{C}[S_4/D_4]$	3	1	0	1	1
$\mathcal{I}_4 \cong V^{(4)}$	1	1	1	1	1
$V^{(2,2)}$	2	0	-1	0	2

As before, $\{\mathbf{D_4} + (\mathbf{1,2}).\mathbf{D_4} + (\mathbf{2,3}).\mathbf{D_4}\}$ spans the trivial representation, and the set $\{\mathbf{D_4} - (\mathbf{1,2}).\mathbf{D_4}, \mathbf{D_4} - (\mathbf{2,3}).\mathbf{D_4}\}$ spans the complementary subspace.

$\mathbb{C}[\mathbf{S_4}/\mathbf{S_3}]$. Using 2-cycles as a transversal for S_4/S_3 makes the computations easier;

$$S_4/S_3 = \{S_3, (1, 4).S_3, (2, 4).S_3, (3, 4).S_3\}.$$

Note that the number of cosets fixed by S_4 in S_4/S_3 is the same as the number of monomials fixed by S_4 in $\mathcal{P}_1(x, y, z, w)$. Subtracting off the trivial character yields the character of a 3-dimensional complementary subspace, which is irreducible by Exercise 7.19, (and equivalent to $V^{(3,1)}$, although at this point it's not obvious why). The trivial representation is spanned by the orbit sum

$$\{\mathbf{S_3} + (\mathbf{1, 4}).\mathbf{S_3} + (\mathbf{2, 4}).\mathbf{S_3} + (\mathbf{3, 4}).\mathbf{S_3}\},$$

and the set $\{\mathbf{S_3} - (\mathbf{1, 4}).\mathbf{S_3},\ \mathbf{S_3} - (\mathbf{2, 4}).\mathbf{S_3},\ \mathbf{S_3} - (\mathbf{3, 4}).\mathbf{S_3}\}$ spans the complementary subspace.

S_4	[1] (1)	[6] (1, 2)	[8] (1, 2, 3)	[6] (1, 2, 3, 4)	[3] (1, 2)(3, 4)
$\mathbb{C}[S_4/S_3]$	4	2	1	0	0
$\mathcal{I}_4 \cong V^{(4)}$	1	1	1	1	1
$\mathcal{W}_4 \cong V^{(3,1)}$	3	1	0	-1	-1

$\mathbb{C}[\mathbf{S_4}/\mathbf{K_4}]$. Since the non-trivial elements of K_4 don't lie in S_3, we can take the elements of S_3 as a transversal;

$$S_4/K_4 = \{K_4, (1, 2).K_4, (1, 3).K_4, (2, 3).K_4, (1, 2, 3).K_4, (1, 3, 2).K_4\}.$$

Taking inner products of characters yields the following character table.

S_4	[1] (1)	[6] (1, 2)	[8] (1, 2, 3)	[6] (1, 2, 3, 4)	[3] (1, 2)(3, 4)
$\mathbb{C}[S_4/K_4]$	6	0	0	0	6
$\mathcal{I}_4 \cong V^{(4)}$	1	1	1	1	1
$\mathcal{A}_4 \cong V^{(1,1,1,1)}$	1	-1	1	-1	1
$V^{(2,2)}$	2	0	-1	0	2
$V^{(2,2)}$	2	0	-1	0	2

$\mathbb{C}[\mathbf{S_4}/\mathbf{R_4}]$. You should have the procedure figured out by now. Inner products of characters shows that the trivial and $(2, 2)$ representations each appear with multiplicity one. Subtracting these from the permutation character of $\mathbb{C}[S_4/R_4]$

yields a new irreducible character, that of $V^{(2,1,1)}$. We'll leave the reader to decide whether or not to work out explicit bases for the representations.

S_4	[1] (1)	[6] (1, 2)	[8] (1, 2, 3)	[6] (1, 2, 3, 4)	[3] (1, 2)(3, 4)
$\mathbb{C}[S_4/R_4]$	6	0	0	2	2
$\mathcal{I}_4 \cong V^{(4)}$	1	1	1	1	1
$V^{(2,2)}$	2	0	-1	0	2
$V^{(2,1,1)}$	3	-1	0	1	-1

$\mathbb{C}[S_4/A_3]$.

S_4	[1] (1)	[6] (1, 2)	[8] (1, 2, 3)	[6] (1, 2, 3, 4)	[3] (1, 2)(3, 4)
$\mathbb{C}[S_4/A_3]$	8	0	2	0	0
$\mathcal{I}_4 \cong V^{(4)}$	1	1	1	1	1
$\mathcal{A}_4 \cong V^{(1,1,1,1)}$	1	-1	1	-1	1
$\mathcal{W}_4 \cong V^{(3,1)}$	3	1	0	-1	-1
$V^{(2,1,1)}$	3	-1	0	1	-1

Observe that along the way we have produced a complete character table for the irreducible representations of S_4.

Character table for S_4.

S_4	[1] (1)	[6] (1, 2)	[8] (1, 2, 3)	[6] (1, 2, 3, 4)	[3] (1, 2)(3, 4)
$\mathcal{I}_4 \cong V^{(4)}$	1	1	1	1	1
$\mathcal{W}_4 \cong V^{(3,1)}$	3	1	0	-1	-1
$V^{(2,2)}$	2	0	-1	0	2
$V^{(2,1,1)}$	3	-1	0	1	-1
$\mathcal{A}_4 \cong V^{(1,1,1,1)}$	1	-1	1	-1	1

Remark 9.69 For the record, also note that the $\chi_{V^{(3,1)}} = \chi_{V^{(2,1,1)}} \chi_{\mathcal{A}_4}$. This is true generally; if λ' is the partition conjugate to λ (Sect. 8.1), then $\chi_{V^{\lambda'}} = \chi_{V^{\lambda}} \chi_{\mathcal{A}_n}$. The usual proof involves tensor products.

Exercise 9.46 We will demonstrate a few computations for $\mathrm{Ind}_{S_2}^{S_3} \mathcal{A}_2$. Choose

$$\{\sigma_1, \sigma_2, \sigma_3\} = \{(1), (1, 3), (2, 3)\}$$

as our transversal for S_3/S_2. Then if $g = (1, 2, 3)$ for example, we have;

$$\sigma_1^{-1} g \sigma_1 = (1)(1, 2, 3)(1) = (1, 2, 3) \notin H, \quad \text{so } \widetilde{\sigma_1^{-1} g \sigma_1} = 0.$$
$$\sigma_2^{-1} g \sigma_1 = (1, 3)(1, 2, 3)(1) = (1, 2) \in H, \quad \text{so } \widetilde{\sigma_2^{-1} g \sigma_1} = \widetilde{(1, 2)} = -1.$$
$$\sigma_3^{-1} g \sigma_1 = (2, 3)(1, 2, 3)(1) = (1, 3) \notin H, \quad \text{so } \widetilde{\sigma_3^{-1} g \sigma_1} = 0.$$
$$\vdots$$
$$\sigma_3^{-1} g \sigma_2 = (2, 3)(1, 2, 3)(1, 2) = (1) \in H, \quad \text{so } \widetilde{\sigma_3^{-1} g \sigma_1} = \widetilde{(1)} = 1$$
$$\sigma_1^{-1} g \sigma_3 = (1)(1, 2, 3)(2, 3) = (1, 2) \in H, \quad \text{so } \widetilde{\sigma_1^{-1} g \sigma_3} = \widetilde{(1, 2)} = -1.$$
$$\vdots$$

Thus,

$$\widetilde{(1, 2, 3)} = \begin{pmatrix} 0 & 0 & -1 \\ -1 & 0 & 0 \\ 0 & 1 & 0 \end{pmatrix}.$$

Since $\text{Ind}_{S_2}^{S_3} \mathcal{A}_2 \cong \mathcal{A}_3 \oplus \mathcal{W}_3$ by Exercise 9.35, and referring to the character table in Sect. 7.1, we have $\chi_{\text{Ind} \mathcal{A}_2}(1, 2, 3) = \chi_{\mathcal{A}_3}(1, 2, 3) + \chi_{\mathcal{W}_3}(1, 2, 3) = 1 + -1 = 0$.

Exercise 9.49 Our goal is to show that the map $\widehat{\varphi}$ is independent of the choice of a transversal for G/H. So suppose that $\{g_i\}_{i=1}^k$ is another transversal of G/H. Then

$$\mathbf{v} = \sigma_1.\mathbf{w_1} + \cdots + \sigma_k.\mathbf{w_k} = g_1.\mathbf{w_1'} + \cdots + g_k.\mathbf{w_k'}, \quad \text{for some } \mathbf{w_i}, \mathbf{w_i'} \in W.$$

Now if $g_i \in \sigma_i H$, then $g_i = \sigma_i h_i$ for some $h_i \in H$. Also, $\sigma_i.\mathbf{w_i} = g_i.\mathbf{w_i'} = \sigma_i h_i.\mathbf{w_i'}$, and so (of course) $\mathbf{w_i} = h_i.\mathbf{w_i'}$.
Then,

$$\begin{aligned}
\widehat{\varphi}(\mathbf{v}) &= \widehat{\varphi}(\sigma_1.\mathbf{w_1} + \cdots + \sigma_k.\mathbf{w_k}) \\
&= \sigma_1.\varphi(\mathbf{w_1}) + \cdots + \sigma_k.\varphi(\mathbf{w_k}) \quad \text{(definition of } \widehat{\varphi}) \\
&= \sigma_1.\varphi(h_1.\mathbf{w_1'}) + \cdots + \sigma_k.\varphi(h_k.\mathbf{w_k'}) \text{ (since } \mathbf{w_i} = h_i \mathbf{w_i'}) \\
&= \sigma_1 h_1.\varphi(\mathbf{w_1'}) + \cdots + \sigma_k h_k.\varphi(\mathbf{w_k'}) \quad \text{(since } \varphi \text{ is } H\text{-linear)} \\
&= g_1.\varphi(\mathbf{w_1'}) + \cdots + g_k.\varphi(\mathbf{w_k'}) \\
&= \widehat{\varphi}(g_1.\mathbf{w_1'} + \cdots + g_k.\mathbf{w_k'}).
\end{aligned}$$

Exercise 9.56 The map $f_{\mathbf{w}} \in \mathcal{F}$ since, for $h \in H$ and $x \in G$, we have:

$$h^{-1}.[f_{\mathbf{w}}(x)] = \begin{cases} h^{-1}.[x^{-1}.\mathbf{w}] = (xh)^{-1}.\mathbf{w} & \text{if } x \in H; \\ 0 & \text{if } x \notin H; \end{cases}$$
$$= f_{\mathbf{w}}(xh). \quad \text{since } xh \in H \Leftrightarrow x \in H.$$

Note: For some the following hints, we will only record the steps where the outcome is non-zero since otherwise the computations are trivial.

Exercise 9.57 The map Ψ is clearly linear since, for $x \in H$ and $\mathbf{w}_1, \mathbf{w}_2 \in W$, we have

$$\begin{aligned}[\Psi(\mathbf{w}_1 + \mathbf{w}_2)](x) &= f_{\mathbf{w}_1 + \mathbf{w}_2}(x) \\ &= x^{-1}.(\mathbf{w}_1 + \mathbf{w}_2) \\ &= x^{-1}.\mathbf{w}_1 + x^{-1}.\mathbf{w}_2 \\ &= f_{\mathbf{w}_1}(x) + f_{\mathbf{w}_2}(x) \\ &= [\Psi(\mathbf{w}_1)](x) + [\Psi(\mathbf{w}_2)](x) \\ &= [\Psi(\mathbf{w}_1) + \Psi(\mathbf{w}_2)](x). \end{aligned}$$

Similarly, $\Psi(r\mathbf{w}) - r\Psi(\mathbf{w})$ for any scalar r.

Now let $h \in H$ and $\mathbf{w} \in W$. Then, for $x \in H$ we have

$$\begin{aligned}[h.\Psi(\mathbf{w})](x) &= [h.f_{\mathbf{w}}](x) & \text{(since the action of } H \text{ is on } \mathcal{F}) \\ &= f_{\mathbf{w}}(h^{-1}x) & \text{(the action of } H < G \text{ on } f_{\mathbf{w}} \in \mathcal{F}) \\ &= (h^{-1}x)^{-1}.\mathbf{w} & \text{(definition of } f_{\mathbf{w}}) \\ &= (x^{-1}h).\mathbf{w} \\ &= (x^{-1}).[h.\mathbf{w}] \\ &= f_{h.\mathbf{w}}(x) \\ &= [\Psi(h.\mathbf{w})](x). \end{aligned}$$

Thus $h.\Psi(\mathbf{w}) = \Psi(h.\mathbf{w})$, so Ψ is an H- linear map. We have again used the fact that, if $h \in H$, then $xh \in H \Leftrightarrow x \in H$.

Exercise 9.58 Let $\mathbf{w}_1, \mathbf{w}_2 \in W$. Then

$$\begin{aligned}\Psi(\mathbf{w}_1) = \Psi(\mathbf{w}_2) &\Leftrightarrow f_{\mathbf{w}_1}(x) = f_{\mathbf{w}_2}(x) & \text{for all } x \in G \\ &\Leftrightarrow x^{-1}.\mathbf{w}_1 = x^{-1}.\mathbf{w}_2 & \text{for all } x \in H < G \\ &\Leftrightarrow \mathbf{w}_1 = \mathbf{w}_2. & \text{(from the definition of group action)} \end{aligned}$$

Therefore Ψ is an H-linear isomorphism from W onto its image $\widehat{W} \subset \mathcal{F}$, so W and $\widehat{W}$ are equivalent as representations of H.

Exercise 9.60 The proof is basically the same as Exercise 9.56, but with more bookkeeping.

Let $h_0 \in H$, and let $x \in G$. Then

$$
h_0^{-1} \cdot f_{g,\mathbf{w}}(x) = \begin{cases} h_0^{-1} \cdot (h^{-1} \cdot \mathbf{w}), & \text{if } x = gh \text{ for some } h \in H; \\ 0, & \text{if } x \notin gH. \end{cases}
$$

$$
= \begin{cases} (hh_0)^{-1} \cdot \mathbf{w}, & \text{if } x = gh \text{ for some } h \in H; \\ 0, & \text{if } x \notin gH. \end{cases}
$$

$$
= \begin{cases} (hh_0)^{-1} \cdot \mathbf{w}, & \text{if } xh_0 = g(hh_0) \text{ for some } h \in H, \\ & (\text{since } x = gh \Rightarrow xh_0 = ghh_0); \\ 0, & \text{if } x \notin gH, \end{cases}
$$

$$
= f_{g,\mathbf{w}}(xh_0), \qquad \text{and hence } f_{g,\mathbf{w}} \in \mathcal{F}.
$$

Exercise 9.61 Let $\sigma, x, g \in G$. Then

$$
[\sigma \cdot f_{g,\mathbf{w}}](x) = f_{g,\mathbf{w}}(\sigma^{-1}x) \quad (\text{the action of } \sigma \text{ on } \mathcal{F})
$$

$$
= \begin{cases} h^{-1} \cdot \mathbf{w}, & \text{if } \sigma^{-1}x = gh \text{ for some } h \in H; \\ 0, & \text{if } \sigma^{-1}x \notin gH. \end{cases}
$$

$$
= \begin{cases} h^{-1} \cdot \mathbf{w}, & \text{if } x = \sigma gh \text{ for some } h \in H; \\ 0, & \text{if } x \notin \sigma gH. \end{cases}
$$

$$
= f_{\sigma g,\mathbf{w}}(x).
$$

Exercise 9.62 Let $f \in \mathcal{F}$. Then for any $x \in G$,

$$
\begin{aligned}
f(x) &= f(\sigma h) & & (\text{for some } \sigma \in G \text{ and some } h \in H, \text{ since } G/H \text{ partitions } G) \\
&= h^{-1} \cdot [f(\sigma)] & & (\text{since } f \in \mathcal{F} \text{ and } f(\sigma) \in W) \\
&= h^{-1} \cdot \mathbf{w} & & (\text{where } \mathbf{w} = f(\sigma)) \\
&= f_{\sigma,\mathbf{w}}(x) & & (\text{since } x = \sigma h, \text{ and by the definition of } f_{\sigma,\mathbf{w}}(x)).
\end{aligned}
$$

Exercise 9.64 The fact that Φ is linear follows from the definition of addition and scalar multiplication of functions. Let e be the identity in G, let $f_1, f_2 \in \mathcal{F}$, and let r be a scalar. Then

$$
\Phi(f_1 + f_2) = [f_1 + f_2](e) = f_1(e) + f_2(e) = \Phi(f_1) + \Phi(f_2), \text{ and } [r\Phi](f) = rf(e) = \Phi(rf).
$$

Now let $h \in H$. Then for every $f \in \mathcal{F}$ we have

$$
\begin{aligned}
h.[\Phi(f)] &= h.[f(e)] \text{ (since the action of } H \text{ is on } W) \\
&= f(eh^{-1}) \text{ (since } f \text{ is in } \mathcal{F}) \\
&= f(h^{-1}e) \\
&= [h.f](e) \text{ (the action of } G \text{ on } \mathcal{F}) \\
&= \Phi(h.f).
\end{aligned}
$$

Whence Φ is H-linear.

Exercise 9.65   For any $\mathbf{w} \in W$ we have

$$
\begin{aligned}
\Phi \circ \Psi(\mathbf{w}) &= \Phi(f_{\mathbf{w}}) \\
&= f_{\mathbf{w}}(e) \\
&= e.\mathbf{w} \\
&= \mathbf{w}.
\end{aligned}
$$

Similarly, for any $f \in \widehat{W}$ and any $x \in H$ we have

$$
\begin{aligned}
[\Psi \circ \Phi(f)](x) &= [\Psi(f(e))](x) \\
&= f_{f(e)}(x) \\
&= x^{-1}.f(e) \qquad \text{(definition of } f_w(x)) \\
&= f(ex) \qquad \text{(since } f \in \mathcal{F} \text{ and } x \in H) \\
&= f(x).
\end{aligned}
$$

Exercise 9.67   Let's start by thinking about what the notation means. Since $\Phi \in \operatorname{Hom}_G(U, \operatorname{Ind} W) = \operatorname{Hom}_G(U, \mathcal{F})$, we have that $\Phi(u) \in \mathcal{F}$, so the notation $\phi(u) = [\Phi(u)](e)$ means to evaluate the function $\Phi(u) : G \to W$ at $e \in G$. It is routine to check that ϕ and Φ are linear, and that the image of ϕ lies in W.

   In what follows it is useful to use explicit notation for the representations (π, U) of G and (ρ, W) of H. Denote by ρ^G the induced action of G on $\mathcal{F}$. Using this notation,

$$
\mathcal{F} = \{f : G \to W \mid \rho(h^{-1})f(x) = f(xh), \ h \in H, \ x \in G\}.
$$

Also, we have

$$
\phi \in \operatorname{Hom}_H(\operatorname{Res} U, W) \quad \text{means that} \quad \phi(\pi(h)u) = \rho(h)\phi(u),
$$

and

$$
\Phi \in \operatorname{Hom}_G(U, \operatorname{Ind} W) \quad \text{means that} \quad \Phi(\pi(g)u) = \rho^G(h)\Phi(u).
$$

First, we check that $[\Phi(u)]$ actually lies in $\mathcal{F}$. For $x \in G$ and $h \in H$;

$$
\begin{aligned}
[\Phi(u)](xh) &= \phi(\pi[xh]^{-1}u) && \text{(definition of } \Phi) \\
&= \phi(\pi[h^{-1}]\pi[x^{-1}]u)) \\
&= \rho(h^{-1})\phi(\pi[x^{-1}]u) && \text{(because } \phi \text{ intertwines } \pi \text{ and } \rho) \\
&= \rho(h^{-1})[\Phi(u)](x). && \text{Hence } \Phi(u) \in \mathcal{F}.
\end{aligned}
$$

To show that Φ is a G-map, let $x, g \in G$. Then;

$$
\begin{aligned}
[\rho^G(g)\Phi(u)](x) &= [\Phi(u)](g^{-1}x) && \text{(the action of } \rho^G \text{ on } \Phi(u) \in \mathcal{F}) \\
&= \phi(\pi[g^{-1}x]^{-1}u) && \text{(definition of } \Phi) \\
&= \phi(\pi[x^{-1}g]u) \\
&= \phi(\pi[x^{-1}]\pi[g]u) \\
&= [\Phi(\pi(g)u)](x). && \text{(definition of } \Phi)
\end{aligned}
$$

To show that ϕ is an H-map;

$$
\begin{aligned}
\phi(\pi(h)u) &= [\Phi(\pi(h)u)](e) && \text{(definition of } \phi) \\
&= [\rho^G(h)\Phi(u)](e) && \text{(since } \Phi \text{ intertwines } \pi \text{ and } \rho^G) \\
&= [\Phi(u)](h^{-1}e) && \text{(the action of } \rho^G \text{ on } \mathcal{F}) \\
&= [\Phi(u)](eh^{-1}) \\
&= \rho(h)[\Phi(u)](e) && \text{(since } \Phi(u) \in \mathcal{F} \text{ and } [\Phi(u)](e) \in W) \\
&= \rho(h)\phi(u)). && \text{(definition of } \phi)
\end{aligned}
$$

Furthermore, these processes are inverses of each other. When we successively apply the definitions, we have that the composition $\phi \to \Phi \to \phi$ is the identity on $\mathrm{Hom}_H(\mathrm{Res}\,U, W)$, and the composition $\Phi \to \phi \to \Phi$ is the identity on $\mathrm{Hom}_G(U, \mathrm{Ind}\,W)$. To wit;

$$
\phi(u) = [\Phi(u)](e) = \phi(\pi[e^{-1}]u) = \phi(u),
$$

and

$$
\begin{aligned}
[\Phi(u)](x) &= \phi[\pi(x^{-1})u] && \text{(definition of } \Phi) \\
&= [\Phi(\pi(x^{-1})u)](e) && \text{(definition of } \phi) \\
&= [\rho^G(x^{-1})\Phi(u)](e) && \text{(since } \Phi\colon U \to \mathcal{F} \text{ is a } G\text{-map)} \\
&= [\Phi(u)](xe) && \text{(the action of } \rho^G \text{ on } \mathcal{F}) \\
&= [\Phi(u)](x).
\end{aligned}
$$

Since the assignments $\phi \to \Phi$ and $\Phi \to \phi$ are linear (you should check this), it follows that $\mathrm{Hom}_H(\mathrm{Res}\,U, W)$ and $\mathrm{Hom}_G(U, \mathrm{Ind}\,W)$ are isomorphic as vector spaces.

Direct Products of Groups, Young Subgroups and Permutation Modules

10

In this chapter we produce new representations of S_n induced from trivial representations of a special class of subgroups of S_n called Young subgroups. Not surprisingly, we will first need to discuss more group theory.

10.1 Direct Products of Groups

Similar to the direct sum of vector spaces (Sect. 2.4), we have the direct product of groups and subgroups.

Definition 10.1 A group G is the *(internal) direct product* of its subgroups H and K if:

(1) Every $g \in G$ can be written as $g = hk$ for some $h \in H$ and $k \in K$.
(2) Every element of H commutes with every element of K.
(3) The subgroups H and K intersect trivially, that is, $H \cap K = \{e\}$.

In this case, we write $G = H \times K$.

We can also form the *external direct product* of two groups H and K by defining $H \times K = \{(h, k) \mid h \in H, k \in K\}$, and where the group operation is componentwise: $(h, k) \times (h', k') = (hh', kk')$. Here the identity in $H \times K$ is the element (e_H, e_K), and we embed H (say) as a subgroup of $H \times K$ as $H = \{(h, e_K) \mid h \in H\}$.

Exercise* 10.2 Show that the two definitions of $H \times K$ are equivalent. Since the distinction is often immaterial or clear from the context, we will usually drop the designation "internal" or "external."

© The Author(s), under exclusive license to Springer Nature Switzerland AG 2022
R. M. Howe, *An Invitation to Representation Theory*, SUMS Readings,
https://doi.org/10.1007/978-3-030-98025-2_10

Exercise* 10.3 Extend this definition to the direct product of a finite number of groups or subgroups, written $\prod_{i=1}^{k} G_i := G_1 \times G_2 \times \cdots \times G_k$. Hint: Sect. 2.4.

Exercise* 10.4 Let $G = H \times K$. Show that $ghg^{-1} \in H$ for all $h \in H$ and all $g \in G$.

Remark 10.5 A subgroup H of a group G with the property that $ghg^{-1} \in H$ for all $h \in H$ and all $g \in G$ is called a *normal subgroup*, and property (2) in the definition is often replaced with the (weaker) condition that both H and K are normal subgroups. Normal subgroups are an important topic in the study of groups, but they are not essential for our purposes.

Exercise* 10.6 Where is property (2) required in the definition? What about the weaker property that H and K are normal?

Example 10.7 Let $(\mathbb{C}^*, \times)$ be the group of non-zero complex numbers under multiplication. Let $\mathbb{S}^1$ denote the subgroup $\{a + bi \mid a^2 + b^2 = 1\}$, and let $\mathbb{R}^+$ denote the subgroup of positive real numbers. Since every complex number z can be written in polar form as $z = r(\cos\theta + i\sin\theta)$, we have $(\mathbb{C}^*, \times) = \mathbb{R}^+ \times \mathbb{S}^1$.

Example 10.8 The direct product of symmetric the groups $S_n \times S_m$ embeds as a subgroup of S_{n+m} in the obvious way, with S_n acting on the first n integers and S_m acting on the last m integers.

Non-Example 10.9 The symmetric group S_3 is not the direct product of its subgroups $S_2 = \{(1), (1, 2)\}$ and $A_3 = \{(1), (1, 2, 3), (1, 3, 2)\}$.

Exercise 10.10 Show that the factorization $g = hk$, for $h \in H$ and $k \in K$ as in Definition 10.1 (1), is unique. Hint: Exercise 2.15.

Exercise* 10.11 Prove the converse to Exercise 10.10; if every $g \in H \times K$ has a unique factorization as $g = hk$ for some $h \in H$ and some $k \in K$, then $H \cap K = \{e\}$.

Exercise* 10.12 If H and K are finite groups, what is the order of $G = H \times K$? Extend this result to the direct product of a finite number of groups.

Exercise* 10.13 Show that if H and K are groups, then $H \times K$ and $K \times H$ are isomorphic as groups.

Exercise 10.14 Show that if H and K are both Abelian groups, then $H \times K$ is Abelian. Show that if either H or K is non-Abelian, then $H \times K$ is non-Abelian. Extend this result to a product of finitely-many groups.

Exercise* 10.15 Which of the properties of Definition 10.1 does the group S_3 from Non-Example 10.9 satisfy, and which properties fail?

10.2   Young Subgroups and Permutation Modules

Consider one of the symmetric groups such as S_{12} acting on the set $\{1, 2, 3, \ldots, 12\}$. It should be clear that those elements of S_{12} that permute only the numbers (say) $\{3, 5, 8\}$ constitute a subgroup of S_{12} that is isomorphic to S_3, and similarly the subgroup that permutes only the numbers $\{1, 2, 9, 11\}$ is isomorphic to S_4. We denote these subgroups by $S_{\{3,5,8\}}$ and $S_{\{1,2,9,11\}}$ respectively. It should take just a minute to check that these two subgroups intersect trivially and commute with each other, and therefore we can form their direct product $S_{\{3,5,8\}} \times S_{\{1,2,9,11\}}$, which is itself a subgroup of S_{12} that is isomorphic to $S_3 \times S_4$.

More generally, define $\mathbf{n} := \{1, 2, 3, \ldots, n\}$, and consider the composition

$$\lambda = (\lambda_1, \lambda_2, \ldots, \lambda_k) \models n.$$

A *dissection* of $\mathbf{n}$ is a set-partition of $\mathbf{n}$ into mutually disjoint subsets $\mathbf{n}_i^\lambda$, with $|\mathbf{n}_i^\lambda| = \lambda_i$. The *Young subgroup* of S_n corresponding to the composition λ, or more explicitly the Young subgroup corresponding to the dissection $\{\mathbf{n}_1^\lambda, \mathbf{n}_2^\lambda, \ldots, \mathbf{n}_k^\lambda\}$, is the direct product

$$S_{\mathbf{n}_1^\lambda} \times S_{\mathbf{n}_2^\lambda} \times \cdots \times S_{\mathbf{n}_k^\lambda}.$$

By the above remarks, this is isomorphic to the subgroup $S_\lambda := S_{\lambda_1} \times S_{\lambda_2} \times \cdots \times S_{\lambda_k}$, and by Exercise 10.13 we can change the order of the subgroups in this direct product and assume that λ is a partition of n.

To continue the example, the dissection $\mathbf{12} = \{3, 5, 8\} \cup \{1, 2, 9, 11\} \cup \{4, 6\} \cup \{7, 10, 12\}$ is indexed by the composition $(3, 4, 2, 3) \models 12$. The Young subgroup corresponding to this dissection is the direct product

Equation 10.16

$$S_{\{3,5,8\}} \times S_{\{1,2,9,11\}} \times S_{\{4,6\}} \times S_{\{7,10,12\}} \cong$$

$$S_3 \times S_4 \times S_2 \times S_3 \cong S_4 \times S_3 \times S_3 \times S_2 = S_{(4,3,3,2)}.$$

If μ and λ are equivalent compositions of n, *i.e.*, the same up to order, then S_μ and S_λ are isomorphic (in fact, they are conjugate) as subgroups of S_n. Similar to Example 10.8, it is usually convenient to think of S_λ as

$$S_{\{1,2,\ldots,\lambda_1\}} \times S_{\{\lambda_1+1,\lambda_1+2,\ldots,\lambda_1+\lambda_2\}} \times \cdots \times S_{\{n-\lambda_k+1,n-\lambda_k+2,\ldots,\lambda_k\}}.$$

Example 10.17 The Young subgroup in Eq. 10.16 is isomorphic to the Young subgroup $S_{\{1,2,3,4\}} \times S_{\{5,6,7\}} \times S_{\{8,9,10\}} \times S_{\{11,12\}}$.

Example 10.18 The group S_2 is the Young subgroup $S_{(2,1)}$ of S_3, and the Young subgroup $S_{(2,1,1)}$ of S_4.

Example 10.19 The entire group S_n, and the trivial subgroup $\{(1)\}$, are Young subgroups of S_n corresponding to $\lambda = (n)$ and $\lambda = (1^n)$ respectively.

Example 10.20 For a Young tableau T, the horizontal subgroup H_T and the vertical subgroup V_T from Sect. 8.3 are Young subgroups.

Non-Example 10.21 For any $n > 2$, the alternating group A_n is not a Young subgroup of S_n since no subsets of $\mathbf{n}$ are fixed by A_n.

Exercise 10.22 For $\lambda = (\lambda_1, \lambda_2, \ldots, \lambda_n) \vdash n$, define $\lambda! := \lambda_1! \lambda_2! \ldots \lambda_n!$ Check that $|S_\lambda| = \lambda!$. Hint: Exercise 10.12.

We next consider the action of S_n on the coset space $\mathbb{C}[S_n/S_\lambda]$, which we recognize from Sect. 9.5 as the representation of S_n induced from the trivial representation of S_λ. Such a representation, *i.e.*, induced from the trivial representation of a Young subgroup S_λ, is an example of a Young permutation module.

Definition 10.23 The representation of S_n induced from the trivial representation of a Young subgroup S_λ is called a *Young permutation module* or just a *permutation module*, traditionally denoted M^λ.

Remark 10.24 The classical construction of M^λ is on a different space. We are breaking with tradition somewhat by calling M^λ any representation induced from the trivial representation of a Young subgroup S_λ, although the notation can vary widely in the literature. See Remark 10.37.

Remark 10.25 Do not confuse Young permutation *modules* with permutation *representations*, as in Remark 1.25. The terminology is another unfortunate historical artifact.

Exercise* 10.26 Show that the dimension of $\mathbb{C}[S_n/S_\lambda] = n!/\lambda!$.

Example 10.27 The permutation module $M^{(n)}$ is induced from the trivial representation of S_n, and therefore is itself the trivial representation of S_n. The permutation module $M^{(1^n)}$ is induced from the trivial representation of the trivial subgroup, and therefore is equivalent to the left regular representation of S_n on $\mathbb{C}[S_n]$.

We will explore several ways to realize Young permutation modules. Here is one that we are already familiar with.

Exercise 10.28 Write down the Young subgroups for S_3. Show that $\mathcal{P}_{(1,0,0)}(x, y, z)$, $\mathcal{P}_{(1,1,0)}(x, y, z)$ and $\mathcal{P}_{(2,1,0)}(x, y, z)$ are permutation modules for S_3. Hint: What monomials are fixed by which Young subgroups?

Exercise 10.29 Choose some Young subgroups of S_n for $n \geq 4$. Which monomials in $\mathcal{P}(x_1, x_2, \ldots, x_n)$ are fixed by these subgroups? What are the corresponding permutation modules?

Exercise 10.30 For $\alpha = (\alpha_1, \alpha_2, \ldots, \alpha_n) \vdash k$ show that $\mathcal{P}_\alpha(x_1, x_2, \ldots, x_n) \subset \mathcal{P}_k$ is a permutation module for S_n. What is the corresponding Young subgroup?

We saw above that $\mathcal{P}_{(1,0,0)}(x, y, z)$ and $\mathcal{P}_{(1,1,0)}(x, y, z)$ are both equivalent to $M^{(2,1,0)}$ as representations of S_3 since the Young subgroup $S_2 = S_{(2,1,0)}$ acts trivially on the one-dimensional subspaces spanned by the the monomials **z** and **xy** respectively.

Similarly, the S_3-invariant subspaces $\mathcal{P}_{(2,2,0)}(x, y, z)$ and $\mathcal{P}_{(3,1,1)}(x, y, z)$ are both equivalent to $M^{(2,1,0)}$ since $S_{(2,1,0)}$ fixes the monomial $\mathbf{x^2 y^2}$, and also fixes the monomial $\mathbf{xyz^3}$.

After working through a few more examples and referring back to Sect. 2.3, we see that what is important is not the **type** of the monomials in the subspace $\mathcal{P}_\alpha$ but their **signature**, as in Remark 2.12. The truth of the following proposition is pretty clear. The hard part is to work out the notation in order to write down a coherent proof. You should give it a try.

Proposition 10.31 *If the polynomial subspace $\mathcal{P}_\alpha$ has signature λ, then both $\mathcal{P}_\alpha$ and $\mathbb{C}[S_n/S_\lambda]$ are equivalent to M^λ as representations of S_n since induced representations are unique up to equivalence by Proposition 9.53.*

Example 10.32 The polynomial space $\mathcal{P}_{(3,1,1,0)}(z, y, z, w)$ has signature $(2, 1, 1)$, and thus is equivalent to $M^{(2,1,1)}$ as a representation of S_4.

Note that permutation modules are, in general, not irreducible. The technique for decomposing permutation modules into irreducibles is discussed in Chap. 12.

10.3 Decomposition of Polynomial Spaces into Permutation Modules

Our eventual goal is to decompose polynomial spaces into irreducible representations of S_n, but we haven't yet developed all of the required tools. As a first step towards this goal, we show how to decompose polynomial spaces into Young permutation modules.

Let S_n act on $\mathcal{P}_k$, the homogeneous polynomials of degree k. Let $\alpha = (\alpha_1, \alpha_2, \ldots, \alpha_n) \vdash k$ of length n, and let $\lambda = (\lambda_1, \lambda_2, \ldots, \lambda_n) \vdash n$. Then the number of invariant subspaces $\mathcal{P}_\alpha$ in $\mathcal{P}_k$ that are equivalent to the permutation module M^λ is equal to the number of partitions $\alpha \vdash k$ with signature $\lambda \vdash n$.

For example, with $k = 4$ and $n = 3$, the partitions of 4 with length 3 are $(4, 0, 0)$, $(3, 1, 0)$, $(2, 2, 0)$, and $(2, 1, 1)$. Consequently, $\mathcal{P}_{(4,0,0)}$, $\mathcal{P}_{(2,1,1)}$, and $\mathcal{P}_{(2,2,0)}$ are the S_3-invariant subspaces of $\mathcal{P}_4(x, y, z)$ equivalent to $M^{(2,1,0)}$; and $\mathcal{P}_{(3,1,0)}$ is the S_3-invariant subspace equivalent to $M^{(1,1,1)}$. It follows that

$$\mathcal{P}_4(x, y, z) = \mathcal{P}_{(4,0,0)} \oplus \mathcal{P}_{(3,1,0)} \oplus \mathcal{P}_{(2,2,0)} \oplus \mathcal{P}_{(2,1,1)} \cong 3\, M^{(2,1,0)} \oplus M^{(1,1,1)}.$$

Note that we require α to have length n, adding trailing zeros as needed.

A more visual way to determine this decomposition is to count the number of ways to fill the Young diagram λ with non-negative integer entries from α so that:

- the entries in each row are equal (since the variables commute),
- the entries in separate rows are all distinct, and
- the sum of all entries is equal to k.

For example, the tableaux below depict the three ways to fill the diagram $\lambda = (2, 1, 0)$ for $k = 4$.

$$\begin{array}{|c|c|} \hline 0 & 0 \\ \hline 4 \\ \cline{1-1} \end{array} \,,\quad \begin{array}{|c|c|} \hline 1 & 1 \\ \hline 2 \\ \cline{1-1} \end{array} \,,\quad \begin{array}{|c|c|} \hline 2 & 2 \\ \hline 0 \\ \cline{1-1} \end{array} .$$

We will see a similar method of counting when we discuss Kostka numbers in Chap. 12.

Exercise* 10.33 Decompose the polynomial space $\mathcal{P}_7(x, y, z, w)$ into permutation modules of S_4.

10.4   More Permutation Modules: Tabloids and Polytabloids

We have seen that the coset space $\mathbb{C}[S_n/S_\lambda]$, and a polynomial space $\mathcal{P}_\alpha(x_1, \ldots, x_n)$ where α has signature λ, are both induced from a trivial representation of the Young subgroup S_λ, and are therefore equivalent to a Young permutation module M^λ as representations of S_n. In this section we demonstrate yet another way to construct such a representation that has theoretical and computational applications. We will also demonstrate S_n-maps between these various incarnations of permutation modules.

A *Young tabloid* is an equivalence class of Young tableaux, two tableaux being equivalent if their rows contain the same entries. To articulate this fact, the diagram for a tabloid is written without the vertical lines, and we denote by $\{T\}$ the tabloid consisting of all tableaux that are equivalent to the tableau T. A *standard Young tabloid*, or just a *standard tabloid*, is a tabloid equivalent to a standard Young tableau. We will see that tabloids are a convenient way to visualize and keep track of Young subgroups and their cosets, as well as a convenient way to keep track of monomials.

Example 10.34 Here are some equivalent tableaux labeled by the tabloid on the right.

$$
T = \begin{array}{|c|c|c|}\hline 1 & 2 & 3 \\\hline 4 & 5 \\\cline{1-2} 6 & 7 \\\cline{1-2} 8 \\\cline{1-1}\end{array} \sim \begin{array}{|c|c|c|}\hline 1 & 3 & 2 \\\hline 5 & 4 \\\cline{1-2} 7 & 6 \\\cline{1-2} 8 \\\cline{1-1}\end{array} \sim \cdots \sim \begin{array}{|c|c|c|}\hline 2 & 1 & 3 \\\hline 4 & 5 \\\cline{1-2} 7 & 6 \\\cline{1-2} 8 \\\cline{1-1}\end{array} \in \begin{array}{ccc} \overline{1 \ \ 2 \ \ 3} \\ 4 \ \ 5 \\ \overline{6 \ \ 7} \\ \overline{8} \end{array} = \{T\}
$$

The symmetric group S_n acts on Young tableaux, and hence on Young tabloids, by permuting the entries in the obvious way, *i.e.*, $\sigma.\{T\} = \{\sigma.T\}$. For example,

$$
(1,4,6).\{T\} = (1,4,6).\begin{array}{cc} \overline{1 \ \ 2 \ \ 3} \\ 4 \ \ 5 \\ \overline{6 \ \ 7} \\ \overline{8} \end{array} = \begin{array}{cc} \overline{4 \ \ 2 \ \ 3} \\ 6 \ \ 5 \\ \overline{1 \ \ 7} \\ \overline{8} \end{array} = \begin{array}{cc} \overline{2 \ \ 3 \ \ 4} \\ 5 \ \ 6 \\ \overline{1 \ \ 7} \\ \overline{8} \end{array}.
$$

We denote by $\mathcal{T}$ the set of tabloids with n entries, and by $\mathcal{T}^\lambda$ (for $\lambda \vdash n$) the set of tabloids with shape λ. We then construct $\mathbb{C}[\mathcal{T}]$, the vector space of *polytabloids*, in a familiar way; we declare that the distinct tabloids are basis vectors, and then take all linear combinations over $\mathbb{C}$. We obtain a representation of S_n on $\mathbb{C}[\mathcal{T}]$ by linearly extending the action of S_n on the basis vectors, and for each $\lambda \vdash n$ we have the S_n-invariant subspace $\mathbb{C}[\mathcal{T}^\lambda]$ of *polytabloids with shape* λ.
For example, if

$$
\mathbf{v} = \begin{array}{cc}\overline{1 \ \ 2}\\3 \ \ 4\end{array} + 2\begin{array}{cc}\overline{2 \ \ 3}\\1 \ \ 4\end{array} + 3\begin{array}{cc}\overline{1 \ \ 3}\\2 \ \ 4\end{array} \in \mathbb{C}[\mathcal{T}^{(2,2)}],
$$

then

$$
(1,2,3).\mathbf{v} = \begin{array}{cc}\overline{2 \ \ 3}\\1 \ \ 4\end{array} + 2\begin{array}{cc}\overline{1 \ \ 3}\\2 \ \ 4\end{array} + 3\begin{array}{cc}\overline{1 \ \ 2}\\3 \ \ 4\end{array}.
$$

Exercise 10.35 Show that the action of S_n on $\mathcal{T}^\lambda$ is transitive.

Now, observe in Example 10.34 that the Young subgroup $S_{(3,2,2,1)}$ acts trivially on the tabloid $\{T\}$. Also notice that we can obtain all possible tabloids with shape $(3, 2, 2, 1)$ by permuting the entries of $\{T\}$.

Remark 10.36 Since polytabloids are vectors, we should be writing them in boldface, but the typesetting software doesn't seem to support boldface tabloids.

Remark 10.37 Regarding Remark 10.24, the notation $\mathbb{C}[\mathcal{T}^\lambda]$ for the space of polytabloids is consistent with such commonly used notation as $\mathbb{C}[S_n]$ for the group algebra, the notation $\mathbb{C}[G/H]$ for the coset space, and the notation $\mathbb{C}[\mathcal{T}_{\lambda\mu}]$ used in Chap. 12. The notation can vary widely in the literature. For example, the reference [FH] uses the notation U_λ for any representation induced from the trivial representation of S_λ. The reference [St] uses M_λ for the set of tabloids with shape λ, and $\mathcal{F}(M_\lambda)$ for the vector space of polytabloids. Some authors use M^λ to denote the representation of S_n on the coset space $\mathbb{C}[S_n/S_\lambda]$.

Exercise 10.38 Work through a few more examples and verify the following statements:

(1) A Young subgroup S_λ acts trivially on some basis vector $\{T\} \in \mathcal{T}^\lambda$.
(2) A basis for the vector space $\mathbb{C}[\mathcal{T}^\lambda]$ consists of all translates of $\{T\}$ by the cosets of S_λ. That is, $\mathbb{C}[\mathcal{T}^\lambda] = \langle\!\langle \{\sigma.\{T\} \mid \sigma S_\lambda \in S_n/S_\lambda\} \rangle\!\rangle$.
(3) The action of of S_n on $\mathbb{C}[\mathcal{T}^\lambda]$ depends only on the left coset of S_λ. *i.e.*, if $\sigma S_\lambda = \sigma' S_\lambda$, then $\sigma.\{T\} = \sigma'.\{T\}$
(4) The dimension of $\mathbb{C}[\mathcal{T}^\lambda]$ is equal to the dimension of $\mathbb{C}[S_n/S_\lambda] = n!/\lambda!$.

Conclude that the representation of S_n on $\mathbb{C}[\mathcal{T}^\lambda]$ is induced from the trivial representation of S_λ, and therefore $\mathbb{C}[\mathcal{T}^\lambda]$ and the Young permutation module M^λ are equivalent representations of S_n.

Exercise* 10.39 Let $\{T\}$ be a Young tabloid with shape λ that is fixed by the Young subgroup S_λ. Show that the map $S_n/S_\lambda \to \mathbb{C}[\mathcal{T}^\lambda]$ defined on the basis vectors by $\sigma S_\lambda \mapsto \sigma.\{T\}$ is a S_n-isomorphism.

We now have equivalent incarnations of the permutation module M^λ: as the coset space $\mathbb{C}[S_n/S_\lambda]$, as the space $\mathcal{P}_\alpha$ of polynomials where α has signature λ, and as the space of polytabloids $\mathbb{C}[\mathcal{T}^\lambda]$. Exercise 10.39 above exhibits an explicit G-isomorphism between $\mathbb{C}[S_n/S_\lambda]$ and $\mathbb{C}[\mathcal{T}^\lambda]$.

We can also construct an S_n-isomorphism between $\mathbb{C}[\mathcal{T}^\lambda]$ and the polynomial space $\mathcal{P}_\alpha$ when α has signature λ (Remark 2.12, see also [P]). Consider the correspondence $\mathbb{C}[\mathcal{T}^\lambda] \to \mathcal{P}_\alpha$ that takes each tabloid $\{T\}$ to the monomial

Equation 10.40

$$m_T = \prod_{r=1}^{n} x_r^{\theta(r)},$$

where $\theta(r) = i - 1$ if r is in the ith row of $\{T\}$. For example, if

$$\{T\} = \begin{array}{|ccc|} \hline 1 & 2 & 3 \\ \hline 4 & 5 \\ \hline 6 \\ \hline \end{array} \quad , \quad \text{then} \quad m_T = x_1^0 x_2^0 x_3^0 x_4^1 x_5^1 x_6^2.$$

It is routine to check that this correspondence is well defined, and intertwines the action of S_n on $\mathcal{T}^\lambda$ with the action of S_n on $\mathcal{P}$. As usual, we then extend the map to all of $\mathbb{C}[\mathcal{T}^\lambda]$ by linearity.

Furthermore, we can expand the construction of Eq. 10.40 in the following way. Note first that if μ has signature λ, then $\mathcal{P}_\mu$ and $\mathbb{C}[\mathcal{T}^\lambda]$ have the same dimension, namely $n!/\lambda!$. If $\mu \models k$ has signature $\lambda \vdash n$, then $\mu = (\mu_1^{\lambda_1}, \ldots, \mu_m^{\lambda_m})$. Label the m rows of the diagram for the tabloid with shape λ by $\mu_1, \ldots, \mu_m$, and fill the rows of the tabloid $\{\lambda\}$ with the corresponding variables. This correspondence also intertwines the action of S_n.

Example 10.41 For $\mathcal{P}_{(3,2,2,1)}(x, y, z, w)$ with signature $\lambda = (2, 1, 1, 0)$ the mono-mial $x^3 y^2 z^2 w$ corresponds to the labeled tabloid

$$\begin{array}{c|cc} 2 & y & z \\ \hline 3 & x \\ \hline 1 & w \\ \hline \end{array} \quad .$$

Similarly for $\mathcal{P}_{(4,1,1,0)}(x, y, z, w)$, which also has signature $\lambda = (2, 1, 1, 0)$, the monomial $x^4 yz$ corresponds to

$$\begin{array}{c|cc} 1 & y & z \\ \hline 4 & x \\ \hline 0 & w \\ \hline \end{array} \quad .$$

As a more general example, the tabloid

$$\begin{array}{c|ccc} 2 & x_1 & x_2 & x_3 \\ \hline 4 & x_4 & x_5 \\ \hline 1 & x_6 & x_7 \\ \hline 0 & x_8 \\ \hline \end{array}$$

corresponds to the monomial $x_1^2 x_2^2 x_3^2 x_4^4 x_5^4 x_6 x_7$.

10.5   Hints and Additional Comments

Exercise 10.2 If every element in H commutes with every element in K, then the map from the external direct product to the internal direct product given by $(h, k) \mapsto hk$ is a group isomorphism.

Exercise 10.3 A group G is the direct product of its subgroups $\{G_1, \ldots, G_k\}$ if

(1) The subgroups generate G. That is, $G = G_1 G_2 \cdots G_k := \{g_1 g_2 \cdots g_k \mid g_i \in G_i\}$.
(2) Every element in G_i commutes with every element in G_j whenever $i \neq j$.
(3) $\left(\prod_{i \neq j} G_i \right) \cap G_j = \{e\}$ for all j.

Exercise 10.4 If $h \in H$ and $g \in G$, then $g = h_1 k_1$ for some $h_1 \in H$ and some $k_1 \in K$. Therefore,

$$ghg^{-1} = h_1 k_1 h k_1^{-1} h_1^{-1} = h_1 k_1 k_1^{-1} h h_1^{-1} = h_1 h h_1^{-1} \in H.$$

Exercise 10.6] For $h_1, h_2 \in H$ and $k_1, k_2 \in K$, we require that the product $h_1 k_1 h_2 k_2$ lies in $HK := \{hk \mid h \in H, k \in K\}$. If every element in H commutes with every element of K, then $h_1 k_1 h_2 k_2 = h_1 h_2 k_1 k_2 \in HK$.

If H and K are normal, then $h_1 k_1 h_2 k_2 = h_1 k_1 h_2 (k_1^{-1} k_1) k_2 = h_1 (k_1 h_2 k_1^{-1}) k_1 k_2$. By normality, $(k_1 h_2 k_1^{-1}) = h_3$ for some $h_3 \in H$. Thus, $h_1 k_1 h_2 k_2 = h_1 h_3 k_1 k_2 \in HK$.

Exercise 10.11 If $g \in H \cap K$ then $g = h = k = he = ek$ for some $h \in H$ and some $k \in K$. Since such a factorization is unique, $h = k = e$.

Exercise 10.12 For each $h \in H$ there are $|K|$ ways to form the pair $g = hk$. Since this portrayal of g is unique, it follows that $|H \times K| = |H||K|$, and that $|\prod_{i=1}^{k} G_i| = \prod_{i=1}^{k} |G_i|$.

Exercise 10.13 The map $(h, k) \mapsto (k, h)$ is a group isomorphism from $H \times K$ to $K \times H$. This generalizes to the direct product of any finite number of groups in any order.

Exercise 10.15 Check that every element of S_2 does not commute with every element of A_3. The other two properties are satisfied.

Exercise 10.26 By Exercises 9.30 and 10.22, $\mathrm{Dim} \, \mathrm{Ind}_{S_\lambda}^{S_n} W = \frac{|S_n|}{|S_\lambda|} \mathrm{Dim} \, W = \frac{n!}{\lambda!} \cdot 1$.

Exercise 10.33 The possible signatures are

$$(4, 0, 0, 0), \quad (3, 1, 0, 0), \quad (2, 2, 0, 0), \quad (2, 1, 1, 0), \quad \text{and} \quad (1, 1, 1, 1).$$

Next, we list all of the possible types $\alpha \vdash 7$ with length 4:

$$(7, 0, 0, 0)$$
$$(6, 1, 0, 0)$$
$$(5, 2, 0, 0)$$
$$(5, 1, 1, 0)$$
$$(4, 3, 0, 0)$$
$$(4, 2, 1, 0)$$
$$(4, 1, 1, 1)$$
$$(3, 3, 1, 0)$$
$$(3, 2, 1, 1)$$
$$(2, 2, 2, 1).$$

To illustrate the viewpoint at the end of Sect. 10.3, we work through the list of signatures. First, we see that there is no way to fill the diagram

$$(4, 0, 0, 0) = \boxed{}\boxed{}\boxed{}\boxed{}$$

with the same entry in each box, and so that the entries sum to seven, so there are no copies of $M^{(4,0,0,0)}$ in $\mathcal{P}_7$.

Next we see that

$$\begin{array}{|c|c|c|}\hline 0 & 0 & 0 \\\hline 7 \\\cline{1-1}\end{array} \quad \text{and} \quad \begin{array}{|c|c|c|}\hline 1 & 1 & 1 \\\hline 4 \\\cline{1-1}\end{array}$$

are the only ways to fill the diagram for $(3, 1, 0, 0)$ from the list, so $\mathcal{P}_{(7,0,0,0)}$ and $\mathcal{P}_{(4,1,1,1)}$ are the subspaces of $\mathcal{P}_7$ equivalent to $M^{(3,1,0,0)}$.

Similarly, we see that there is no way to fill the diagram

$$(2, 2, 0, 0) = \begin{array}{|c|c|}\hline & \\\hline & \\\hline\end{array},$$

and only the following types can fill the diagram $(2, 1, 1, 0)$ as prescribed:

$$\begin{array}{|c|c|}\hline 0 & 0 \\\hline 6 \\\cline{1-1} 1 \\\cline{1-1}\end{array}, \quad \begin{array}{|c|c|}\hline 0 & 0 \\\hline 5 \\\cline{1-1} 2 \\\cline{1-1}\end{array}, \quad \begin{array}{|c|c|}\hline 1 & 1 \\\hline 5 \\\cline{1-1} 0 \\\cline{1-1}\end{array}, \quad \begin{array}{|c|c|}\hline 0 & 0 \\\hline 4 \\\cline{1-1} 3 \\\cline{1-1}\end{array}, \quad \begin{array}{|c|c|}\hline 3 & 3 \\\hline 1 \\\cline{1-1} 0 \\\cline{1-1}\end{array}, \quad \text{and} \quad \begin{array}{|c|c|}\hline 1 & 1 \\\hline 3 \\\cline{1-1} 2 \\\cline{1-1}\end{array}.$$

Therefore $\mathcal{P}_{(6,1,0,0)}$, $\mathcal{P}_{(5,2,0,0)}$, $\mathcal{P}_{(5,1,1,0)}$, $\mathcal{P}_{(4,3,0,0)}$, $\mathcal{P}_{(3,3,1,0)}$ and $\mathcal{P}_{(3,2,1,1)}$ are the subspaces of $\mathcal{P}_7(x, y, z, w)$ equivalent to $M^{(2,1,1,0)}$ as representations of S_4.

Finally, we see that

4
2
1
0

is the only way to fill the diagram for $(1, 1, 1, 1)$. Summarizing, we have

$$\mathcal{P}_7(x, y, z, w) \cong 2M^{(3,1,0,0)} \oplus 6M^{(2,1,1,0)} \oplus M^{(1,1,1,1)}.$$

Exercise 10.39 Define $\phi : \mathbb{C}[S_\lambda/S_n] \rightarrow \mathbb{C}[\mathcal{T}^\lambda]$ on the basis vectors $\mathcal{T}^\lambda$ by $\phi(\sigma S_\lambda) = \sigma\{T\}$ and extend by linearity.

The map ϕ is well defined on cosets. If $\sigma S_\lambda = \sigma' S_\lambda$, then $\sigma^{-1}\sigma' \in S_\lambda$. Therefore $\sigma^{-1}\sigma'.\{T\} = \{T\}$ and so $\sigma.\{T\} = \sigma'.\{T\}$.

Now suppose $\phi(\sigma' S_\lambda) = \phi(\sigma S_\lambda)$. Then $\sigma'.\{T\} = \sigma\{T\}$, so that $\sigma^{-1}\sigma'.\{T\} = \{T\}$, and therefore $\sigma^{-1}\sigma' \in S_\lambda$. Consequently, $\sigma' S_\lambda = \sigma S_\lambda$, and hence ϕ is one-to-one.

It is routine to check that ϕ is an S_n-map that is onto $\mathbb{C}[\mathcal{T}^\lambda]$ since S_n acts transitively on the basis vectors $\mathcal{T}^\lambda$.

Specht Modules

11

In this chapter we produce an irreducible representation $\mathcal{S}^\lambda$, called a Specht module, that appears in each Young permutation module $\mathbb{C}[\mathcal{T}^\lambda]$ with multiplicity one, and show that the collection $\{\mathcal{S}^\lambda \mid \lambda \vdash n\}$ is a complete set of pairwise inequivalent irreducible representations of S_n. We demonstrate an explicit basis for $\mathcal{S}^\lambda$, which we then use to obtain bases for irreducible representations in polynomial spaces.

11.1 Construction of Specht Modules

We start by designating a special class of polytabloids. Let T be a Young tableau. From Sect. 8.3, recall the *vertical subgroup*,

$$V_T := \{v \in S_n \mid v \text{ setwise preserves each column of } T\},$$

and the *column anti-symmetrizer*,

$$\mathcal{V}_T := \sum_{v \in V_T} \text{sgn}(v)v \in \mathbb{C}[S_n].$$

Also recall from Exercise 8.14 that $\mathcal{V}_T$ is a relative idempotent.

Definition 11.1 If T is a Young tableau with shape λ, define the *associated polytabloid* by

$$\mathbf{e}_T := \mathcal{V}_T.\{T\} \in \mathbb{C}[\mathcal{T}^\lambda].$$

A *standard polytabloid* is a polytabloid associated to a standard Young tableau. If $\{T'\}$ is one of the basis tabloids in the sum for $\mathbf{e}_T$, we say that $\{T'\}$ *appears* in $\mathbf{e}_T$.

© The Author(s), under exclusive license to Springer Nature Switzerland AG 2022
R. M. Howe, *An Invitation to Representation Theory*, SUMS Readings,
https://doi.org/10.1007/978-3-030-98025-2_11

Example 11.2 For the tableau $T = \begin{array}{|c|c|c|} \hline 1 & 2 & 3 \\ \hline 4 & 5 \\ \cline{1-2} \end{array}$ we have

$$\mathcal{V}_T = (1) - (1,4) - (2,5) + (1,4)(2,5),$$

and

$$\mathbf{e}_T = \frac{\overline{\begin{array}{ccc} 1 & 2 & 3 \\ 4 & 5 \end{array}}}{} - \frac{\overline{\begin{array}{ccc} 2 & 3 & 4 \\ 1 & 5 \end{array}}}{} - \frac{\overline{\begin{array}{ccc} 1 & 3 & 5 \\ 2 & 4 \end{array}}}{} + \frac{\overline{\begin{array}{ccc} 3 & 4 & 5 \\ 1 & 2 \end{array}}}{}.$$

Exercise 11.3 Confirm that that each tabloid appearing in any $\mathbf{e}_T$ has coefficient ± 1.

Exercise 11.4 Let

$$T_1 = \begin{array}{|c|c|} \hline 1 & 2 \\ \hline 3 & 4 \\ \hline \end{array}, \quad \text{and let} \quad T_2 = \begin{array}{|c|c|} \hline 2 & 1 \\ \hline 3 & 4 \\ \hline \end{array}.$$

Observe that $\{T_1\} = \{T_2\}$, but that $\mathbf{e}_{T_1} \neq \mathbf{e}_{T_2}$. Conclude that a polytabloid $\mathbf{e}_T$ depends on the tableau T, and not just on the tabloid $\{T\}$.

Exercise 11.5 Write out more examples of associated and standard polytabloids.

Exercise* 11.6 Show that if $\sigma \in S_n$, then $\mathcal{V}_{\sigma.T} = \sigma \mathcal{V}_T \sigma^{-1}$.

Definition 11.7 For each $\lambda \vdash n$, define the *Specht module*

$$\mathcal{S}^\lambda := \text{span of } \{\mathbf{e}_T \mid T \text{ has shape } \lambda\}.$$

Exercise* 11.8 Show that for $\sigma \in S_n$, $\sigma.\mathbf{e}_T = \mathbf{e}_{\sigma.T}$. Conclude that $\mathcal{S}^\lambda$ is a *cyclic* S_n-*module* generated by $\mathbf{e}_T$, where T is any tableau with shape λ. In other words, for any tableau T with shape λ, the Specht module $\mathcal{S}^\lambda = $ span of $\{\sigma.\mathbf{e}_T \mid \sigma \in S_n\}$. Also conclude that $\mathcal{S}^\lambda$ is a left-invariant $\mathbb{C}[S_n]$-module.

Exercise 11.9 Determine the Specht modules for S_3 corresponding to $\lambda = (3)$, and $\lambda = (1,1,1)$.

11.2   Irreducibility of Specht Modules

We start with an exercise that illustrates the lemma that follows.

Exercise* 11.10 Let T and V_T be as in Example 11.2. Compute $V_T.\{T_i\}$ for the tabloids

$$T_1 = \frac{\overline{\begin{array}{ccc} 1 & 3 & 5 \\ 2 & 4 \end{array}}}{}, \quad T_2 = \frac{\overline{\begin{array}{ccc} 3 & 4 & 5 \\ 1 & 2 \end{array}}}{}, \quad T_3 = \frac{\overline{\begin{array}{ccc} 1 & 3 & 4 \\ 2 & 5 \end{array}}}{},$$

$$T_4 = \frac{\overline{\begin{array}{ccc} 1 & 2 & 4 \\ 3 & 5 \end{array}}}{}, \quad T_5 = \frac{\overline{\begin{array}{cccc} 1 & 2 & 4 & 5 \\ 3 \end{array}}}{}.$$

Lemma 11.11 *Let T and T' be λ-tableaux. Then either $V_T.\{T'\} = \pm e_T$, or $V_T.\{T'\} = 0$.*

Proof The proof uses similar techniques as in the proof of Proposition 8.25. First, note that $T' = \sigma.T$ for some $\sigma \in S_n$ since the action of S_n on the set of λ-tableaux is transitive (Exercise 10.35). By Proposition 8.27, there are only two cases to consider.

Case 1: If $\sigma \in V_T H_T$, then we claim that $V_T.\{T'\} = \pm e_T$.

$$\begin{aligned}
V_T.\{T'\} &= V_T.\{\sigma.T\} && \text{(since } T' = \sigma.T) \\
&= V_T.\{vh.T\} && (\sigma = vh, \text{ for } v \in V_T \text{ and } h \in H_T) \\
&= V_T v.\{h.T\} && \text{(the action of } S_n \text{ on tabloids)} \\
&= V_T v.\{T\} && \text{(since } \{h.T\} = \{T\}) \\
&= \text{sgn}(v)V_T.\{T\} && \text{(by Exercise 8.12)} \\
&= \pm e_T.
\end{aligned}$$

Case 2: If $\sigma \notin V_T H_T$, then there is a pair i, j appearing in both the same row of T' and the same column of T, and therefore the transposition $(i, j) \in V_T \cap H_{T'}$. By Exercise 8.12, $V_T(i, j) = \text{sgn}(i, j)V_T = -V_T$. Hence,

$$\begin{aligned}
V_T.\{T'\} &= V_T(i, j).\{T'\} \text{ (since } (i, j) \in H_{T'}, \text{ so } (i, j).\{T'\} = \{T'\}) \\
&= -V_T\{T'\},
\end{aligned}$$

and so $V_T.\{T'\} = 0$.

$\square$

Corollary 11.12 *If T is a λ-tableau, if T' is λ'-tableau, and if $\lambda' > \lambda$ in the lexicographic order, then $V_T.\{T'\} = 0$.*

Proof As in the proof of Proposition 8.29, the hypothesis $\lambda' > \lambda$ guarantees that there is a pair i, j that appear in the same row of T' and the same column of T. The proof then proceeds as in the proof of Lemma 11.11, Case 2.

$\square$

Lemma 11.13 *If* $\mathbf{w}$ *is a non-zero element in* $\mathcal{S}^{\lambda}$, *then* $\mathcal{V}_T.\mathbf{w} = c\,\mathbf{e}_T$ *for some constant* $c \in \mathbb{C}$.

Proof If $\mathbf{w} \in \mathcal{S}^{\lambda}$, then $\mathbf{w} = \sum_i c_i\{T_i\}$ for some constants c_i. Therefore, by Lemma 11.11,

$$\mathcal{V}_T.\mathbf{w} = \mathcal{V}_T.\sum_i c_i\{T_i\} = \sum_i c_i\mathcal{V}_T.\{T_i\} = \sum_i \pm c_i\mathbf{e}_T = c\,\mathbf{e}_T.$$

$\square$

Proposition 11.14 *The Specht modules* $\mathcal{S}^{\lambda}$ *are irreducible.*

Proof Let $\mathcal{S}^{\lambda} :=$ span of $\{\mathbf{e}_T \mid T \text{ has shape } \lambda\}$, let W be a non-zero subrepresentation of S_n contained in $\mathcal{S}^{\lambda}$, and suppose that $\mathbf{w}$ is a non-zero element in W. Since W is invariant, $\mathcal{V}_T.\mathbf{w} \in W$. By the previous lemma, $\mathcal{V}_T.\mathbf{w} = c\mathbf{e}_T \in W$, and thus $\mathbf{e}_T \in W$. Again, since W is invariant and by Exercise 11.8, $g.\mathbf{e}_T = \mathbf{e}_{g.T} \in W$, and so $\mathcal{S}^{\lambda} \subset W$. Therefore $\mathcal{S}^{\lambda} = W$. $\square$

11.3 Inequivalence of Specht Modules

In order to show that the Specht modules are pair-wise inequivalent, we show that if there is an S_n-isomorphism between $\mathcal{S}^{\lambda}$ and $\mathcal{S}^{\lambda'}$, then $\lambda = \lambda'$.

Proposition 11.15 *Let* $\phi\colon \mathbb{C}[\mathcal{T}^{\lambda}] \to \mathbb{C}[\mathcal{T}^{\lambda'}]$ *be an* S_n-map. *If* $\lambda' > \lambda$ *in the lexicographic ordering, then* $\mathcal{S}^{\lambda} \in \operatorname{Ker}\phi$.

Proof Let T be a standard Young tableau with shape λ, so that $\mathbf{e}_T$ is a generator of $\mathcal{S}^{\lambda}$. Now, for some scalars c_i, $\phi(\{T\}) = c_1\{T'_1\} + \cdots + c_k\{T'_k\}$ is a polytabloid in $\mathbb{C}[\mathcal{T}^{\lambda'}]$. Therefore,

$$
\begin{aligned}
\phi(\mathbf{e}_T) &= \phi(\mathcal{V}_T\{T\}) \\
&= \mathcal{V}_T\phi(\{T\}) \\
&= \mathcal{V}_T(c_1\{T'_1\} + \cdots + c_k\{T'_k\}) \\
&= c_1\mathcal{V}_T\{T'_1\} + \cdots + c_k\mathcal{V}_T\{T'_k\} \\
&= 0
\end{aligned}
$$

since each $\mathcal{V}_T\{T'_i\} = 0$ by Corollary 11.12. $\square$

Corollary 11.16 *If* $\phi\colon \mathbb{C}[\mathcal{T}^{\lambda}] \to \mathbb{C}[\mathcal{T}^{\lambda'}]$ *is an* S-map, *and if* $\mathcal{S}^{\lambda} \notin \operatorname{Ker}\phi$, *then* $\lambda' \leq \lambda$.

Proposition 11.17 *The Specht modules are pairwise inequivalent. Consequently, the collection*

$$\{\mathcal{S}^\lambda \mid \lambda \vdash n\}$$

is a complete set of inequivalent irreducible representations of S_n.

Proof Let λ and λ' be partitions of n. Since no $\mathbf{e}_T$ is zero for any tableau T, the Specht modules are all nonzero.

By Maschke's Theorem, there are complementary submodules W and W' so that $\mathbb{C}[\mathcal{T}^\lambda] = \mathcal{S}^\lambda \oplus W$, and $\mathbb{C}[\mathcal{T}^{\lambda'}] = \mathcal{S}^{\lambda'} \oplus W'$. Let P_λ be the projection from $\mathbb{C}[\mathcal{T}^\lambda]$ onto $\mathcal{S}^\lambda$, and let $i_{\lambda'}$ be the inclusion of $\mathcal{S}^{\lambda'}$ into $\mathbb{C}[\mathcal{T}^{\lambda'}]$. If $\mathcal{S}^\lambda$ and $\mathcal{S}^{\lambda'}$ are equivalent, then there is a non-zero S_n-isomorphism $\theta\colon \mathcal{S}^\lambda \to \mathcal{S}^{\lambda'}$. Now consider the composition $\phi\colon \mathbb{C}[\mathcal{T}^\lambda] \to \mathbb{C}[\mathcal{T}^{\lambda'}]$ given by[1]

$$\phi\colon \mathbb{C}[\mathcal{T}^\lambda] \overset{P_\lambda}{\twoheadrightarrow} \mathcal{S}^\lambda \overset{\theta}{\to} \mathcal{S}^{\lambda'} \overset{i_{\lambda'}}{\hookrightarrow} \mathbb{C}[\mathcal{T}^{\lambda'}].$$

Then $\phi\colon \mathbb{C}[\mathcal{T}^\lambda] \to \mathbb{C}[\mathcal{T}^{\lambda'}]$ is a non-zero S_n-map that does not contain $\mathcal{S}^\lambda$ in its kernel. By Corollary 11.16, we have $\lambda' \leq \lambda$. By a similar argument, $\lambda \leq \lambda'$, so $\lambda = \lambda'$.

Since, by the results of Chaps. 7 and 8, all of the the inequivalent irreducible representations are in one-to-one correspondence with the integer partitions of n, we have a complete set of irreducible representations of S_n.   □

Exercise* 11.18 Use the alternative version of the Young symmetrizer, $\mathbf{E}_T^* := \mathcal{V}_T \mathcal{H}_T$ in Remark 8.6, and the techniques from Sect. 8.5, to obtain another proof that the Specht modules are a complete set of inequivalent irreducible representations of S_n.

Corollary 11.19 *The Specht module $\mathcal{S}^\lambda \in \mathbb{C}[\mathcal{T}^\lambda]$, and the module $V^\lambda \in \mathbb{C}[S_n]$ from Definition 8.21, are equivalent irreducible representations of S_n.*

11.4   The Standard Basis for Specht Modules

The goal of this section is to demonstrate the following theorem, which will be a consequence of Propositions 11.30 and 11.55.

---

[1] The notation $\twoheadrightarrow$ indicates a surjective (onto) mapping, and the notation $\hookrightarrow$ indicates an injective (one-to-one) mapping.

Theorem 11.20 *The set*

$$\mathcal{B}^\lambda = \{\mathbf{e}_T \mid T \text{ is a standard Young tableau with shape } \lambda\}$$

is a basis for $\mathcal{S}^\lambda$.

Corollary 11.21 *The dimension of the Specht module $\mathcal{S}^\lambda$ is the number of standard Young tableaux with shape λ, which is often denoted in the literature by the symbol f^λ.*

11.4.1 Linear Independence

Here is a simple exercise that illustrates an important idea.

Exercise 11.22 Let

$$T_1 = \begin{array}{|c|c|}\hline 1 & 2 \\\hline 3 \\\cline{1-1}\end{array} \quad \text{and} \quad T_2 = \begin{array}{|c|c|}\hline 1 & 3 \\\hline 2 \\\cline{1-1}\end{array}.$$

Show that the polytabloids $\mathbf{e}_{T_1}$ and $\mathbf{e}_{T_2}$ are linearly independent.

In order to prove that, in general, the set $\mathcal{B}^\lambda$ is linearly independent, we start by defining a total order on tabloids.

Remark 11.23 Here and in what follows, the symbol $<$ is used to denote several different orderings: on real numbers, partitions, tabloids, tableaux, and column-equivalent tableaux. It should be clear from the context which ordering we mean.

There are several ways to order or partially order tableaux, tabloids, etc. Using a convenient ordering can considerably streamline a proof.

Definition 11.24 Define a total order on the set of tabloids with shape λ by saying that $\{T_1\} < \{T_2\}$ if the largest entry in a different row of $\{T_1\}$ than $\{T_2\}$ appears in a higher row of $\{T_1\}$ than of $\{T_2\}$ (see [JK], Section 7.2).

For example,

$$\begin{array}{ccc}\hline 3 & 4 & 5 \\\hline 1 & 2 \\\cline{1-2}\end{array} \;<\; \begin{array}{ccc}\hline 2 & 4 & 5 \\\hline 1 & 3 \\\cline{1-2}\end{array} \quad \text{and} \quad \begin{array}{ccc}\hline 3 & 4 & 5 \\\hline 1 & 2 \\\cline{1-2}\end{array} \;<\; \begin{array}{ccc}\hline 1 & 2 & 3 \\\hline 4 & 5 \\\cline{1-2}\end{array}.$$

Exercise* 11.25 Order the standard tabloids with shape $(3, 2)$.

Exercise* 11.26 Show that the ordering of Definition 11.24 is, in fact, a total order.

Exercise 11.27 For the standard tableau

$$T = \begin{array}{|c|c|c|} \hline 1 & 2 & 3 \\ \hline 4 & 5 \\ \cline{1-2} \end{array},$$

determine $v.\{T\}$ for some $v \in V_T$. Repeat this for several other choices of v, and for several other standard tableaux.

The previous exercise illustrates the following lemma.

Lemma 11.28 *If T is a standard Young tableau, and if $v \in V_T$, with $v \neq (1)$, then $\{v.T\} < \{T\}$ in the ordering of Definition 11.24.*

Proof The largest entry in $\{T\}$ moved by v is moved up, and it therefore appears in a higher row.   □

Lemma 11.28 now yields the following.

Proposition 11.29 *Let T be a standard Young tableau. Then $\{T\}$ is the maximum tabloid appearing in $\mathbf{e}_T$.*

Proposition 11.30 *The set $\mathcal{B}^\lambda$ of standard polytabloids with shape λ is linearly independent.*

Proof We begin by ordering the standard Young tabloids with shape λ,

$$\{T_1\} > \{T_2\} > \cdots > \{T_k\},$$

and consider the relation $\sum c_i \mathbf{e}_{T_i} = 0$ for some scalars c_i.

Now, for each standard tableau T_i and each $v \in V_{T_i}$, each of the tabloids $\{vT_i\}$ appearing in $\mathbf{e}_{T_i}$ have coefficient ± 1 (Exercise 11.3). Since all the standard tabloids are distinct, there is no way to cancel the maximum standard tabloid $\{T_1\}$ appearing with non-zero coefficient in the relation $\sum c_i \mathbf{e}_{T_i} = 0$, which forces $c_1 = 0$. Repeating this reasoning inductively, we conclude that all of the $c_i = 0$, and hence the set $\mathcal{B}^\lambda$ is linearly independent.   □

Exercise 11.31 You should work through more examples, say with $\lambda = (2, 2)$ or $\lambda = (3, 2)$, and verify that the standard polytabloids are linearly independent. In particular, verify the assertion that there is no way to cancel the maximum tabloid appearing in each standard polytabloid.

11.4.2 Span

We are left with the somewhat messier task of establishing that the standard polytabloids in $\mathbb{C}[\mathcal{T}^\lambda]$ span $\mathcal{S}^\lambda$. That is, if T is a Young tableau with associated polytabloid $\mathbf{e}_T$, then $\mathbf{e}_T$ can be written as a linear combination of standard polytabloids. We start with an exercise.

Exercise* 11.32 Consider the standard Young tableaux

$$T_1 = \begin{array}{|c|c|} \hline 1 & 2 \\ \hline 3 & 4 \\ \hline \end{array} \quad \text{and} \quad T_2 = \begin{array}{|c|c|} \hline 1 & 3 \\ \hline 2 & 4 \\ \hline \end{array}.$$

For the non-standard tableaux

$$T_3 = \begin{array}{|c|c|} \hline 1 & 4 \\ \hline 3 & 2 \\ \hline \end{array} \quad \text{and} \quad T_4 = \begin{array}{|c|c|} \hline 1 & 2 \\ \hline 4 & 3 \\ \hline \end{array},$$

write the polytabloids $\mathbf{e}_{T_3}$ and $\mathbf{e}_{T_4}$ as a linear combination of the standard polytabloids $\mathbf{e}_{T_1}$ and $\mathbf{e}_{T_2}$.

Such an ad hoc approach won't work in general, so we start by defining yet another order that refines the lexicographic order on partitions to obtain a total order on Young tableaux. We can then use a procedure referred to as a straightening algorithm and employ induction.

Definition 11.33 If T is a Young tableau, then the *column word* of T consists of the entries of T read top-to-bottom and left-to-right.

For example, the column word of

$$\begin{array}{|c|c|c|} \hline 1 & 2 & 3 \\ \hline 4 & 5 \\ \cline{1-2} 6 & 7 \\ \cline{1-2} 8 \\ \cline{1-1} \end{array} \quad \text{is} \quad (1, 4, 6, 8, 2, 5, 7, 3).$$

Definition 11.34 Define a total order on the set of all Young tableaux by saying that $T' > T$ if either:

(1) the shape of T' is larger than the shape of T in the lexicographic order, or
(2) the tableaux T' and T have the same shape, and the largest entry that is in a different box in the two numberings occurs earlier in the column word of T' than in the column word of T.

For example,

$$
\begin{array}{|c|c|c|}
\hline 3 & 2 & 4 \\
\hline 6 & 7 \\
\cline{1-2}
1 & 5 \\
\cline{1-2}
8 \\
\cline{1-1}
\end{array}
\quad > \quad
\begin{array}{|c|c|c|}
\hline 1 & 2 & 3 \\
\hline 4 & 5 \\
\cline{1-2}
6 & 7 \\
\cline{1-2}
8 \\
\cline{1-1}
\end{array}
$$

because the largest entry that is in a different box is 7, which appears in the column word $(3, 6, 1, 8, 2, 7, 5, 4)$ before it appears in the column word $(1, 4, 6, 8, 2, 5, 7, 3)$.

Exercise* 11.35 Order the standard Young tableaux with shape $(3, 2)$. It is helpful to first identify the largest entry that is in a different box.

Exercise* 11.36 Verify that the ordering of Definition 11.34 is, in fact, a total order.

Remark 11.37 There are other definitions of "column word," as well as other ways to totally order or partially order tableaux and tabloids, that can prove useful in obtaining further results. The ordering in Definition 11.34 follows [P]. See also [Sa], Section 2.5, and [JK], Section 7.2.

11.4.3 A Straightening Algorithm

By using this procedure, we can write any polytabloid as a linear combination of polytabloids that are "smaller" in the ordering of Definition 11.33, then employ induction.

First we will provide a few examples that illustrate the procedure, and then prove that it works in general. We've placed the computations for these examples in a table that the reader should refer to along the way.

To start, we may assume that the entries of T are increasing down the columns since, if not, there is an element $v \in V_T$ such that the tableau $S = v.T$ has increasing columns. But by Exercises 8.12 and 11.8, along with the definition $e_T := V_T.\{T\}$, we have that $e_S = e_{v.T} = v.e_T = \text{sgn}(v)e_T$, so e_T is a linear combination of standard polytabloids whenever e_S is.

Definition 11.38 A *row descent* in a tableau T is a pair of entries in the same row i and adjacent columns j, $j + 1$ such that $T_{i,j} > T_{i,j+1}$. A *column descent* is defined analogously.

Example 11.39 Consider the Young tableau

$$T_1 = \begin{array}{|c|c|} \hline 1 & 2 \\ \hline 4 & 3 \\ \hline \end{array}$$

which has a descent $4 > 3$ in the bottom row, and thus is not standard. Our goal is to write $\mathbf{e}_{T_1}$ as a linear combination of standard polytabloids using the straightening algorithm.

We begin by locating the decent $4 > 3$, taking all entries below and including the entry 4 and place them in a set A, and taking all entries above and including the entry 3 and place them in a set B. We next re-partition the disjoint union $A \uplus B$ into new subsets A' and B' so that

- $A' \uplus B' = A \uplus B$,
- $|A'| = |A|$ and $|B'| = |B|$.

For each possible re-partition of $A \uplus B$, we create a new tableau T_k by replacing A with A' and B with B' in T_1, and in such a way that the entries are increasing down the columns.

A	B		
4	2, 3	$T_1 =$	$\begin{array}{\|c\|c\|} \hline 1 & 2 \\ \hline 4 & 3 \\ \hline \end{array}$

A'	B'		
2	3, 4	$T_2 =$	$\begin{array}{\|c\|c\|} \hline 1 & 3 \\ \hline 2 & 4 \\ \hline \end{array}$
3	2, 4	$T_3 =$	$\begin{array}{\|c\|c\|} \hline 1 & 2 \\ \hline 3 & 4 \\ \hline \end{array}$

Now consider the subgroup $S_{A \uplus B} = S_{\{2,3,4\}}$ of S_4, its subgroup $S_A \times S_B = \{(1)\} \times \{(1), (2,3)\} = \{(1), (2,3)\}$, and the cosets $\pi'(S_A \times S_B)$ in $S_{A \uplus B}/(S_A \times S_B)$:

$$\begin{aligned} (S_A \times S_B) &= \{(1), (2,3)\}, \\ (2,4)(S_A \times S_B) &= \{(2,4), (2,4)(2,3)\} = \{(2,4), (2,3,4)\}, \\ (3,4)(S_A \times S_B) &= \{(3,4), (3,4)(2,3)\} = \{(3,4), (2,4,3)\}. \end{aligned}$$

Check that each element π' in any given coset has the property that $A' = \pi'(A)$ and $B' = \pi'(B)$, but that there is exactly one permutation π_k in each coset that transforms T_1 into a new tableaux $T_k = \pi_k T_1$ in such a way that the columns of T_k are increasing.

In the table below we've recorded the permutations π', as well as the permutations π_k that transform T_1 into these new tableaux $T_k = \pi_k T_1$. For this case the new tableaux T_2 and T_3 are both standard. Verify that, since there is a unique way to order the sets A' and B' so that the resulting tableau T_k has increasing columns, the permutation π_k in the rightmost column of each row in our table is unique.

Exercise 11.40 Verify that both $T_2 < T_1$ and $T_3 < T_1$ in the ordering of Definition 11.34.

A	B	$A' = \pi'(A),\, B' = \pi'(B)$	$T_k = \pi_k.T_1$			
4	2, 3	$\pi' = (1),\, (2,3)$	$T_1 = $	$\begin{array}{cc} 1 & 2 \\ 4 & 3 \end{array}$	$= (1).T_1$	
A'	B'					
2	3, 4	$\pi' = (2,4),\, (2,3,4)$	$T_2 = $	$\begin{array}{cc} 1 & 3 \\ 2 & 4 \end{array}$	$= (2,3,4).T_1$	
3	2, 4	$\pi' = (3,4),\, (2,4,3)$	$T_3 = $	$\begin{array}{cc} 1 & 2 \\ 3 & 4 \end{array}$	$= (3,4).T_1$	

Finally, we form the *Garnir element* $g_{A.B}$ using the permutations π_k;

$$g_{A.B} = \sum_k \operatorname{sgn}(\pi_k)\pi_k = (1) + (2,3,4) - (3,4),$$

and Proposition 11.50 (below) says that

$$g_{A.B}.e_{T_1} = e_{T_1} + e_{T_2} - e_{T_3} = 0.$$

Hence $e_{T_1} = -e_{T_2} + e_{T_3}$ is a linear combination of standard polytabloids.

Exercise* 11.41 Verify that $g_{A.B}.e_{T_1} = 0$ in the above example.

Example 11.42 Here is a slightly more complicated example using the tableau

$$T_1 = \begin{array}{ccc} 2 & 1 & 3 \\ 4 & 5 & \end{array},$$

and tabulating the manipulations as before. Note that T_1 has a descent $2 > 1$ starting in the first column.

A	B	$A' = \pi'(A),\ B' = \pi'(B)$	$T_k = \pi_k . T_1$
2, 4	1	$\pi' = (1),(2,4)$	$T_1 = \begin{array}{\|c\|c\|c\|} \hline 2 & 1 & 3 \\ \hline 4 & 5 \\ \cline{1-2} \end{array} = (1).T_1$
A'	B'		
1, 4	2	$\pi' = (1,2),(1,2,4)$	$T_2 = \begin{array}{\|c\|c\|c\|} \hline 1 & 2 & 3 \\ \hline 4 & 5 \\ \cline{1-2} \end{array} = (1,2).T_1$
1, 2	4	$\pi' = (1,4),(1,4,2)$	$T_3 = \begin{array}{\|c\|c\|c\|} \hline 1 & 4 & 3 \\ \hline 2 & 5 \\ \cline{1-2} \end{array} = (1,4,2).T_1$

Exercise 11.43 Again, verify that both $T_2 < T_1$ and $T_3 < T_1$ in the ordering of Definition 11.34.

In this case we have

$g_{A,B} = (1) - (1,2) + (1,4,2)$, and $g_{A,B}.e_{T_1} = e_{T_1} - e_{T_2} + e_{T_3} = 0$. (Check this).

Thus e_{T_1} is a linear combination of polytabloids e_{T_2} and e_{T_3}. But T_3 has a row descent $4 > 3$ and thus is not standard, so we repeat the straightening algorithm with T_3.

A	B	$A' = \pi'(A),\ B' = \pi'(B)$	$T_k = \pi_k . T_3$
4, 5	3	$\pi' = (1),\ldots$	$T_3 = \begin{array}{\|c\|c\|c\|} \hline 1 & 4 & 3 \\ \hline 2 & 5 \\ \cline{1-2} \end{array} = (1).T_3$
A'	B'		
3, 5	4	$\pi' = (3,4),\ldots$	$T_4 = \begin{array}{\|c\|c\|c\|} \hline 1 & 3 & 4 \\ \hline 2 & 5 \\ \cline{1-2} \end{array} = (3,4).T_3$
3, 4	5	$\pi' = (3,5),\ldots$	$T_5 = \begin{array}{\|c\|c\|c\|} \hline 1 & 3 & 5 \\ \hline 2 & 4 \\ \cline{1-2} \end{array} = (3,5,4).T_3$

For this step we have

$$g_{A,B} = (1) - (3,4) + (3,5,4) \text{ and, } g_{A,B}.e_{T_3} = e_{T_3} - e_{T_4} + e_{T_5} = 0,$$

with e_{T_4} and e_{T_5} both standard polytabloids. Substituting, we obtain $e_{T_1} = e_{T_2} - e_{T_4} + e_{T_5}$, so that e_{T_1} is a linear combination of standard polytabloids. Again, note that both $T_4 < T_3$ and $T_5 < T_3$ in the ordering of Definition 11.34.

Exercise 11.44 Write out the cosets in $S_{A \uplus B}/(S_A \times S_B)$ for the two tables in Example 11.42. Verify that $\mathbf{e}_{T_1} = \mathbf{e}_{T_2} - \mathbf{e}_{T_4} + \mathbf{e}_{T_5}$.

Here is an exercise to check your comprehension.

Exercise* 11.45 Consider the tableau with descent $5 > 4$ in the second row,

$$T_1 = \begin{array}{|c|c|c|} \hline 1 & 2 & 3 \\ \hline 5 & 4 \\ \cline{1-2} 6 \\ \cline{1-1} \end{array}.$$

(1) What are the sets A and B?
(2) What is the subgroup $S_{A \uplus B}$ of S_6? What is $|S_{A \uplus B}|$?
(3) What is the subgroup $S_A \times S_B$ of $S_{A \uplus B}$? Why is this, in fact, a subgroup? What is $|S_A \times S_B|$?
(4) Write out a table similar to the above examples. In particular, determine the permutations π' so that $A' = \pi'(A)$ and $B' = \pi'(B)$. Also determine the permutation π_k that transform the tableau T_1 into a tableau $T_k = \pi_k T_1$ in such a way that T_k has increasing columns.
(5) Write out the cosets $S_{A \uplus B}/(S_A \times S_B)$. Verify that the permutations π' and π_k in each row of the table are in the same coset, and that any permutation in the same coset will transform A to the same A' and B to the same B'.
(6) Verify that, for the new tableaux T_k thus obtained, $T_k < T_1$ using the ordering of Definition 11.34 .
(7) For each k, let π_k be the permutation for which $T_k = \pi_k T_1$ has increasing columns. Verify that the collection $\Pi = \{\pi_k\}$ constitutes a transversal of $S_{A \uplus B}/(S_A \times S_B)$.
(8) What is the Garnir element for this example? Why do we want to form the Garnir element with the transversal Π from (7), and not with some other transversal?
(9) Which of the T_k are standard? Write the non-standard polytabloids as a linear combination of standard polytabloids.
(10) Write $\mathbf{e}_{T_1}$ as a linear combination of standard polytabloids.

Exercise* 11.46 How many ways are there to partition the set $A \uplus B$ into $A' \uplus B'$, with $|A| = |A'|$ and $|B| = |B'|$? Hint: Binomial coefficients. What is the significance of this number?

Exercise* 11.47 Write out a definition for the Garnir element.

For what follows, it will be convenient to generalize the construction used for the column anti-symmetrizer $\mathcal{V}_T$.

Definition 11.48 Given any subset of K of S_n, define the sum

$$K^- := \sum_{k \in K} \text{sgn}(k)k \in \mathbb{C}[S_n].$$

With this notation we have $\mathcal{V}_T = V_T^-$, and the Garnir element $g_{A,B} = \Pi^-$. Note that K need not be a subgroup.

Exercise* 11.49 This is a version of the *Sign Lemma* 2.4.1 of [Sa] that extends the results of Exercise 8.12. Let H be a subgroup of S_n and prove the following:

(1) If $h \in H$, then

$$hH^- = H^-h = \text{sgn}(h)H^-.$$

(2) If the transposition $(b, c) \in H$, then we can factor

$$H^- = s[(1) - (b, c)], \quad \text{for some } s \in \mathbb{C}[S_n].$$

   Hint: Let K be the subgroup $\{(1), (b, c)\}$, and write H as the disjoint union of it K-cosets.

(3) If T is a Young tableau with b, c in the same row of T, and if the transposition $(b, c) \in H$, then

$$H^-\{T\} = 0.$$

   Hint: By hypothesis, $(b, c)\{T\} = \{T\}$, then use (2).

(4) Generalize (2) above. If K is a subgroup of H, and if $\Pi = \{\pi_k\}$ is a transversal of H/K, then $H^- = \Pi^- K^-$.

   We now have the tools to prove the general result.

Proposition 11.50 *Let T, A and B be as in the definition of a Garnir element. If $|A \uplus B|$ is greater than the length of column j of T, (the column whose entries include the elements of A) then $g_{A,B}.e_T = 0$.*

Proof We present the proof as a series of exercises.

Exercise* 11.51 Show that $S_{A \uplus B}^- e_T = 0$. Hint: Use the pigeonhole principal and part (3) of Exercise 11.49 on each $v.\{T\}$ appearing in e_T.

Exercise* 11.52 Show that $g_{A,B}(S_A \times S_B)^- e_T = 0$. Hint: Write $S_{A \uplus B}$ as the disjoint union of $(S_A \times S_B)$-cosets, then use Exercise 11.49 (4) and Definition 11.48 to factor $S_{A \uplus B}^-$. Now use the results of Exercise 11.51.

Exercise* 11.53 Show that $(S_A \times S_B)^- e_T = |S_A \times S_B| e_T$. Hint: Note that $S_A \times S_B \subset V_T$, then apply part (1) of Exercise 11.49 to each $\sigma \in S_A \times S_B$.

Exercise* 11.54 Show that $g_{A.B} e_T = 0$. Hint: String together the results of the previous exercises.

$\square$

Proposition 11.55 *The set $\mathcal{B}^\lambda$ spans $\mathcal{S}^\lambda$.*

Proof Consider the polytabloid e_T. If T is standard we are finished, so suppose that T is not standard. By induction, we may assume that $e_{T'}$ can be written as a linear combination of standard polytabloids whenever $T' < T$ in the ordering of Definition 11.34. Note that we are assuming nothing when there is no T' with $T' < T$, so there is no need to establish a base case for the induction.

Since T is not standard, some adjacent pair of columns, say the jth and $(j+1)$st, have entries $a_1 < a_2 < \cdots < a_r$ and $b_1 < b_2 < \cdots < b_s$, with $a_q > b_q$ for some q. The situation for the jth and $(j+1)$st columns of T looks like this:

$$
\begin{array}{cc}
a_1 & b_1 \\
\wedge & \wedge \\
a_2 & b_2 \\
\wedge & \wedge \\
\cdot & \cdot \\
\cdot & \cdot \\
\cdot & \cdot \\
\wedge & \wedge \\
a_q & > \; b_q \\
\wedge & \wedge \\
\cdot & \cdot \\
\cdot & \cdot \\
\cdot & \wedge \\
\cdot & b_s \\
\cdot & \\
\wedge & \\
a_r. &
\end{array}
$$

So $A = \{a_q, a_{q+1}, \cdots, a_r\}$, $B = \{b_1, b_2, \cdots, b_q\}$, with $b_1 < b_2 < \cdots < b_q < a_q < \cdots < a_r$. Since $|A \uplus B| = 1 +$ (the length of column j), Proposition 11.50 applies, and e_T can be written as a linear combination of the polytabloids e_{T_k} as in the above examples.

Now, the largest entry in T moved by any $\pi_k \in \Pi$ will be from the set A in column j of T and placed in column $j+1$ of $T_k = \pi_k T$. Therefore it will appear later in the column word of T_k, and thus $T_k < T$. By the induction hypothesis, each of the $\mathbf{e}_{T_k}$ can be written as a linear combination of standard polytabloids, and thus $\mathbf{e}_T$ can be written as a linear combination of standard polytabloids, completing the proof.   □

We can now state a partial result towards our goal of decomposing Young permutation modules into S_n-irreducible subspaces.

Proposition 11.56

$$M^\mu \cong S^\mu \oplus \bigoplus_{\lambda > \mu} K_{\lambda\mu} S^\lambda.$$

The multiplicities $K_{\lambda\mu}$ are called Kostka numbers, and we will see a method for determining their value in Chap. 12. The Kostka numbers also appear in other related settings.

Proof $K_{\mu\mu} = 1$ follows from Lemmas 11.11 and 11.13. By Proposition 11.15, $K_{\lambda\mu} = 0$ for $\lambda < \mu$.   □

Example 11.57 By now these examples should be familiar.

$$M^{(1,1,1)} \cong \mathcal{P}_{(2,1,0)} \cong S^{(1,1,1)} \oplus 2\,S^{(2,1,0)} \oplus S^{(3,0,0)} \cong \mathcal{I} \oplus 2\mathcal{W} \oplus \mathcal{A}.$$

$$M^{(2,1,0)} \cong \mathcal{P}_{(1,0,0)} \cong S^{(2,1,0)} \oplus S^{(3,0,0)} \cong \mathcal{W} \oplus \mathcal{I}.$$

11.5   Application to Polynomial Spaces

Consider the polynomial space $\mathcal{P}_{(1,1,0,0)}(\mathbf{x}, \mathbf{y}, \mathbf{z}, \mathbf{w}) = \langle\!\langle \mathbf{xy}, \mathbf{xz}, \dots, \mathbf{zw} \rangle\!\rangle$. From Sect. 10.2 we recognize this as equivalent to the permutation module $M^{(2,2)} \cong \mathbb{C}[\mathcal{T}^{(2,2)}]$ as a representation of S_4. The two standard Young tableaux with shape $(2, 2)$ are

$$T_1 = \begin{array}{|c|c|} \hline 1 & 2 \\ \hline 3 & 4 \\ \hline \end{array} \quad \text{and} \quad T_2 = \begin{array}{|c|c|} \hline 1 & 3 \\ \hline 2 & 4 \\ \hline \end{array},$$

with corresponding standard polytabloids

$$\mathbf{e}_{T_1} = \begin{array}{|c|c|} \hline 1 & 2 \\ \hline 3 & 4 \\ \hline \end{array} - \begin{array}{|c|c|} \hline 2 & 3 \\ \hline 1 & 4 \\ \hline \end{array} - \begin{array}{|c|c|} \hline 1 & 4 \\ \hline 2 & 3 \\ \hline \end{array} + \begin{array}{|c|c|} \hline 3 & 4 \\ \hline 1 & 2 \\ \hline \end{array},$$

$$\mathbf{e}_{T_2} = \frac{\boxed{\begin{array}{cc} 1 & 3 \\ 2 & 4 \end{array}}}{} - \frac{\boxed{\begin{array}{cc} 2 & 3 \\ 1 & 4 \end{array}}}{} - \frac{\boxed{\begin{array}{cc} 1 & 4 \\ 2 & 3 \end{array}}}{} + \frac{\boxed{\begin{array}{cc} 2 & 4 \\ 1 & 3 \end{array}}}{},$$

which is a basis for the Specht module $\mathcal{S}^{(2,2)}$ in the space of polytabloids $\mathbb{C}[\mathcal{T}^{(2,2)}]$. Applying the assignment 10.40 from Sect. 10.4, we have

$$p_1 = \mathbf{zw} - \mathbf{xw} - \mathbf{yz} + \mathbf{xy}, \quad \text{and} \quad p_2 = \mathbf{yw} - \mathbf{xw} - \mathbf{xy} + \mathbf{xz}$$

as basis polynomials for a copy of the Specht module $\mathcal{S}^{(2,2)}$ in $\mathcal{P}_{(1,1,0,0)}(\mathbf{x}, \mathbf{y}, \mathbf{z}, \mathbf{w})$ $\cong M^{(2,2)}$.

Generalizing this construction as in Example 10.41, and applying it to $\mathcal{P}_{(2,2,1,1)}(\mathbf{x}_1, \mathbf{x}_2, \mathbf{x}_3, \mathbf{x}_4)$, we have

$$q_1 = \mathbf{x}_4^2 \mathbf{x}_3^2 \mathbf{x}_1 \mathbf{x}_2 - \mathbf{x}_1^2 \mathbf{x}_4^2 \mathbf{x}_2 \mathbf{x}_3 - \mathbf{x}_2^2 \mathbf{x}_3^2 \mathbf{x}_1 \mathbf{x}_4 + \mathbf{x}_1^2 \mathbf{x}_2^2 \mathbf{x}_3 \mathbf{x}_4,$$

and

$$q_2 = \mathbf{x}_2^2 \mathbf{x}_4^2 \mathbf{x}_1 \mathbf{x}_3 - \mathbf{x}_1^2 \mathbf{x}_4^2 \mathbf{x}_2 \mathbf{x}_3 - \mathbf{x}_2^2 \mathbf{x}_3^2 \mathbf{x}_1 \mathbf{x}_4 + \mathbf{x}_1^2 \mathbf{x}_3^2 \mathbf{x}_2 \mathbf{x}_4$$

as basis polynomials for a copy of the Specht module $\mathcal{S}^{(2,2)}$ in $\mathcal{P}_{(2,2,1,1)}$ $(\mathbf{x}_1, \mathbf{x}_2, \mathbf{x}_3, \mathbf{x}_4) \cong M^{(2,2)}$.

Exercise 11.58 Find basis polynomials for irreducible representations in some other polynomial spaces.

Remark * *11.59* When we apply Eq. 10.40 from Sect. 10.4 to a standard polytabloid $\mathbf{e}_T$, the result is known as a *Specht polynomial*. The discussion following Eq. 10.40 generalizes this construction to other polynomial spaces with the same signature.

11.6   Hints and Additional Comments

Exercise 11.6 By Exercise 8.10, $V_{\sigma.T} = \sigma V_T \sigma^{-1}$, so

$$
\begin{aligned}
V_{\sigma.T} &= \sum_{\tau \in V_{\sigma.T}} \operatorname{sgn}(\tau)\tau \\
&= \sum_{\pi \in V_T} \operatorname{sgn}(\sigma \pi \sigma^{-1}) \sigma \pi \sigma^{-1} \\
&= \sigma \left[\sum_{\pi \in V_T} \operatorname{sgn}(\pi)\pi \right] \sigma^{-1} \\
&= \sigma V_T \sigma^{-1}.
\end{aligned}
$$

Exercise 11.8 Using Exercise 11.6, and the fact that the action of S_n on the set of λ-tableaux is transitive,

$$\mathbf{e}_{\sigma.T} = V_{\sigma.T}\{\sigma.T\} = \sigma V_T \sigma^{-1}.\{\sigma.T\} = \sigma V_T.\{\sigma^{-1}\sigma.T\} = \sigma V_T.\{T\} = \sigma.\mathbf{e}_T.$$

Exercise 11.10

$$\mathcal{V}_T.\{T_1\} = -\mathbf{e}_T, \quad \mathcal{V}_T.\{T_2\} = \mathbf{e}_T, \quad \mathcal{V}_T.\{T_3\} = 0, \quad \mathcal{V}_T.\{T_4\} = 0, \quad \mathcal{V}_T.\{T_5\} = 0.$$

Exercise 11.25

$$
\begin{array}{cc}
\underline{1 \quad 3 \quad 5} \\
\underline{2 \quad 4}
\end{array}
<
\begin{array}{cc}
\underline{1 \quad 2 \quad 5} \\
\underline{3 \quad 4}
\end{array}
<
\begin{array}{cc}
\underline{1 \quad 3 \quad 4} \\
\underline{2 \quad 5}
\end{array}
<
\begin{array}{cc}
\underline{1 \quad 2 \quad 4} \\
\underline{3 \quad 5}
\end{array}
<
\begin{array}{cc}
\underline{1 \quad 2 \quad 3} \\
\underline{4 \quad 5}
\end{array}
$$

Exercise 11.26 If $\{T_1\} \neq \{T_2\}$, then there must be in entry in a different row of $\{T_1\}$ than of $\{T_2\}$, and so there must be a largest such entry. Hence either $\{T_1\} < \{T_2\}$ or $\{T_2\} < \{T_1\}$.

Exercise 11.18 Since $h.\{T\} = \{T\}$ for all $h \in H_T$, we have

$$\mathbf{E}_T^*.\{T\} = \mathcal{V}_T \mathcal{H}_T.\{T\} = \mathcal{V}_T.|H_T|\{T\} = |H_T|\mathbf{e}_T.$$

By Theorem 8.22, distinct partitions of n produce Young symmetrizers that generate distinct inequivalent irreducible representations of S_n.

Exercise 11.32 Note that T_1 and T_3 differ only by a column permutation, so $\mathbf{e}_{T_3} = -\mathbf{e}_{T_1}$. The polytabloid $\mathbf{e}_{T_4} = \mathbf{e}_{T_1} - \mathbf{e}_{T_2}$.

Exercise 11.35

$$
\begin{array}{|c|c|c|}
\hline 1 & 3 & 5 \\ \hline 2 & 4 \\ \cline{1-2}
\end{array}
<
\begin{array}{|c|c|c|}
\hline 1 & 2 & 5 \\ \hline 3 & 4 \\ \cline{1-2}
\end{array}
<
\begin{array}{|c|c|c|}
\hline 1 & 3 & 4 \\ \hline 2 & 5 \\ \cline{1-2}
\end{array}
<
\begin{array}{|c|c|c|}
\hline 1 & 2 & 4 \\ \hline 3 & 5 \\ \cline{1-2}
\end{array}
<
\begin{array}{|c|c|c|}
\hline 1 & 2 & 3 \\ \hline 4 & 5 \\ \cline{1-2}
\end{array}
$$

Exercise 11.36 If $T_1 \neq T_2$ have the same shape, then there must be at least one entry in T_1 that occupies a different position than it occupies in T_2. There must be a largest such entry that consequently appears in a different position in the column word of T_1 than of T_2.

Exercise 11.41 This is mostly an exercise in working carefully.

Exercise 11.45 As before, we've recorded most of the answers in a table. Note that when we are performing these manipulations with any Young tableau, the sets A and B are disjoint. Consequently $S_A \cap S_B = (1)$, the cycles in S_A are disjoint from those in S_B, and hence every element in S_A commutes with every element in S_B. Thus $S_A \times S_B$ satisfies Definition 10.1 for the direct product of groups, and therefore $S_A \times S_B$ is a subgroup of $S_{A \uplus B}$. It follows that $|S_A \times S_B| = |S_A| \times |S_B| = |A|! \times |B|!$, and $|S_{A \uplus B}| = |A \uplus B|! = (|A| + |B|)!$.

We list some of the cosets of $(S_A \times S_B)$ in $S_{A \uplus B}$.

$$\pi_1(S_A \times S_B) = \qquad (S_A \times S_B) \quad = \{(1), (5,6), (2,4), (5,6)(2,4)\}$$
$$\pi_2(S_A \times S_B) = (2,5)(S_A \times S_B) = \{(2,5), (2,5,6), (2,4,5), (2,4,5,6)\}$$
$$\pi_3(S_A \times S_B) = (4,6,5)(S_A \times S_B) = \{(4,6), (4,6,5), (2,6,4), (2,6,5,4)\}$$

etc.

A	B	$A' = \pi'(A),\ B' = \pi'(B)$	$T_k = \pi_k.T_1$
5, 6	2, 4	$\pi' = (1)$	$T_1 = \begin{array}{ccc} 1 & 2 & 3 \\ 5 & 4 \\ 6 \end{array} = (1).T_1$
A'	B'		
2, 6	5, 4	$\pi' = (2,5),\ (2,5,6),\ (2,4,5),$ $(2,4,5,6)$	$T_2 = \begin{array}{ccc} 1 & 5 & 3 \\ 2 & 4 \\ 6 \end{array} = (2,5).T_1$
4, 5	2, 6	$\pi' = (4,6),\ (4,6,5),\ (2,6,4),$ $(2,6,5,4)$	$T_3 = \begin{array}{ccc} 1 & 2 & 3 \\ 4 & 6 \\ 5 \end{array} = (4,6,5).T_1$
2, 5	4, 6	$\pi' = (2,6),\ \ldots$	$T_4 = \begin{array}{ccc} 1 & 4 & 3 \\ 2 & 6 \\ 5 \end{array} = (2,4,6,5).T_1$
4, 6	2, 5	$\pi' = (4,5),\ \ldots$	$T_5 = \begin{array}{ccc} 1 & 2 & 3 \\ 4 & 5 \\ 6 \end{array} = (4,5).T_1$
2, 4	5, 6	$\pi' = (2,5)(4,6),\ (2,6)(4,5),\ \ldots$	$T_6 = \begin{array}{ccc} 1 & 5 & 3 \\ 2 & 6 \\ 4 \end{array} = (2,5)(4,6).T_1$

Note that $T_1 > T_k$ for each $k = 2, \ldots, 6$, and that this would be the case no matter the choice of coset representative π'.

Recall that our goal is to write e_{T_1} as a linear combination of STANDARD polytabloids, so we want to choose as a transversal those permutations π_k that result in each $T_k = \pi_k.T_1$ having increasing columns. Here is an example that gives a feel for what is happening. Consider the row containing T_4 in the above table. If we just applied the permutation $\pi' = (2, 6)$ to the initial tableau

$$T_1 = \begin{array}{|c|c|c|} \hline 1 & 2 & 3 \\ \hline 5 & 4 \\ \cline{1-2} 6 \\ \cline{1-1} \end{array}, \quad \text{we would obtain the tableau} \quad \begin{array}{|c|c|c|} \hline 1 & 6 & 3 \\ \hline 5 & 4 \\ \cline{1-2} 2 \\ \cline{1-1} \end{array},$$

which has descents in the first and second columns. But if we first permute the columns by an element in $S_A \times S_B$,

$$(5, 6)(2, 4) \begin{array}{|c|c|c|} \hline 1 & 2 & 3 \\ \hline 5 & 4 \\ \cline{1-2} 6 \\ \cline{1-1} \end{array} = \begin{array}{|c|c|c|} \hline 1 & 4 & 3 \\ \hline 6 & 2 \\ \cline{1-2} 5 \\ \cline{1-1} \end{array},$$

and then apply the permutation $\pi' = (2, 6)$, we obtain the desired tableau $T_4 = \pi_4 T_1$ with increasing columns. That is, $\pi_4 = (2, 6)(5, 6)(2, 4) \in \pi'(S_A \times S_B)$.

The Garnir element in this example is

$$g_{A,B} = (1) - (2, 5) + (4, 6, 5) - (2, 4, 6, 5) - (4, 5) + (2, 5)(4, 6),$$

and thus

$$\mathbf{e}_{T_1} = \mathbf{e}_{T_2} - \mathbf{e}_{T_3} + \mathbf{e}_{T_4} + \mathbf{e}_{T_5} - \mathbf{e}_{T_6}.$$

Note that we need to repeat the straightening algorithm with the non-standard tableaux T_2, T_4, and T_6. We leave this to the reader. Better yet, since T_2, T_4, and T_6 are all less than T_1, by induction we can assume that their associated polytabloids can be written as a linear combination of standard polytabloids.

Exercise 11.46 Recall that the binomial coefficient $\binom{n}{k} = \frac{n!}{k!(n-k)!}$ gives the number of ways to choose k unordered outcomes from n possibilities.

Applying this to our case, the number of ways to repartition $A \uplus B$ into $A' \uplus B'$ is the number of ways to choose $|A|$ elements from $|A \uplus B|$ possibilities. But $|A \uplus B| = |A| + |B|$, and therefore the number of ways to partition $A \uplus B$ into $A' \uplus B'$

$$= \frac{|A \uplus B|!}{|A|!(|A \uplus B| - |A|)!} = \frac{|A \uplus B|!}{|A|!|B|!} = \frac{|S_{A \uplus B}|}{|S_A||S_B|} = |S_{A \uplus B}/(S_A \times S_B)|$$

$$= \text{ the index } [S_{A \uplus B} : S_A \times S_B].$$

In other words, the number of ways to partition $A \uplus B$ into $A' \uplus B'$ is the number of cosets of $S_A \times S_B$ in $S_{A \uplus B}$.

Since for each coset $\pi(S_A \times S_B)$ there is a unique π_k such that the tableau $T_k = \pi_k.T_1$ has increasing columns, the collection $\Pi = \{\pi_k\}$ is a transversal for $S_{A \uplus B}/(S_A \times S_B)$.

Exercise 11.47 Here is the most general definition:

Let A and B be disjoint sets (usually of positive integers), and let S_A, S_B, and $S_{A \uplus B}$ denote the group of permutations of the sets A, B, and $A \uplus B$ respectively. Choose a transversal $\{\pi_k\}$ of $(S_A \times S_B)/S_{A \uplus B}$. Then a *Garnir element* for A and B is the sum

$$g_{A,B} = \sum \operatorname{sgn}(\pi_k)\pi_k.$$

Exercise 11.49

(1) This is essentially the same as Exercise 8.12.

(2) $H = \biguplus h_i K$, so $H^- = \sum \operatorname{sgn}(h_i)h_i[(1) - (b,c)] = s[(1) - (b,c)]$.

(3) Because $(b,c)\{T\} = \{T\}$ we have

$$H^-\{T\} = s[(1) - (b,c)]\{T\} = s[\{T\} - \{T\}] = 0.$$

Exercise 11.51 To show that $S^-_{A \uplus B}\mathbf{e}_T = 0$, let $v \in V_T$. By the hypothesis that $|A \uplus B|$ is greater than the length of column j along with the pigeonhole principal, there must be a pair $a \in A$ and $b \in B$ that lie in in the same row of $v.T$. Since the transposition (a,b) is an element in $S_{A \uplus B}$, and by part (3) of Exercise 11.49, we have $S^-_{A \uplus B}v.\{T\} - 0$. Since this is true for each $v.\{T\}$ appearing in $\mathbf{e}_T = V_T.\{T\}$, we have $S^-_{A \uplus B}.\mathbf{e}_T = 0$.

Exercise 11.52 To show that $g_{A,B}(S_A \times S_B)^-\mathbf{e}_T = 0$, we write $S_{A \uplus B}$ as the disjoint union of $(S_A \times S_B)$-cosets,

$$S_{A \uplus B} = \biguplus_{\pi_k \in \Pi} \pi_k(S_A \times S_B),$$

which, by Exercise 11.49 (4) and Definition 11.48, gives the factorization

$$S^-_{A \uplus B} = \Pi^-(S_A \times S_B)^- = g_{A,B}(S_A \times S_B)^-.$$

Substituting this into the results of Exercise 11.51, we have

$$g_{A,B}(S_A \times S_B)^-.\mathbf{e}_T = S^-_{A \uplus B}.\mathbf{e}_T = 0.$$

Exercise 11.53 Applying part (1) of Exercise 11.49 to each $\sigma \in S_A \times S_B$, we obtain

$$\operatorname{sgn}(\sigma)\sigma.\mathbf{e}_T = \operatorname{sgn}(\sigma)\sigma.V_T^-\{T\} = V_T^-\{T\} = V_T\{T\} = \mathbf{e}_T,$$

and therefore

$$(S_A \times S_B)^-.\mathbf{e}_T = |S_A \times S_B|\mathbf{e}_T.$$

Exercise 11.54 Stringing together the previous results, we obtain

$$0 = S^-_{A \uplus B} . \mathbf{e}_T = g_{A.B}(S_A \times S_B)^- . \mathbf{e}_T = g_{A.B} . |S_A \times S_B| \mathbf{e}_T = |S_A \times S_B| g_{A.B} . \mathbf{e}_T.$$

Since $|S_A \times S_B| \neq 0$, we must have $g_{A.B} \mathbf{e}_T = 0$

Remark 11.59 Specht polynomials are defined as a product of Vandermonde polynomials (Exercise 5.11). Let T be a Young tableau, and denote by $[T_1], \ldots, [T_k]$ the columns of T. For example, if

$$T = \begin{array}{|c|c|c|} \hline 1 & 2 & 3 \\ \hline 5 & 4 \\ \cline{1-2} 6 \\ \cline{1-1} \end{array}, \quad \text{then } [T_1] = \begin{array}{|c|} \hline 1 \\ \hline 5 \\ \hline 6 \\ \hline \end{array}, \quad [T_2] = \begin{array}{|c|} \hline 2 \\ \hline 4 \\ \hline \end{array} \quad \text{and} \quad [T_3] = \begin{array}{|c|} \hline 3 \\ \hline \end{array}.$$

Let $\Delta([T_j])$ be the Vandermonde polynomial subscripted by the entries in the jth column of T. In this example

$$\Delta([T_1]) = \begin{vmatrix} 1 & 1 & 1 \\ x_1 & x_5 & x_6 \\ x_1^2 & x_5^2 & x_6^2 \end{vmatrix}, \quad \text{and} \quad \Delta([T_2]) = \begin{vmatrix} 1 & 1 \\ x_2 & x_4 \end{vmatrix}.$$

Finally, define the *Specht polynomial*

$$\Delta(T) := \Delta([T_1])\Delta([T_2]) \cdots \Delta([T_k]).$$

It is shown in [Sp] that the set $\{\Delta(T) \mid T \text{ has shape } \lambda\}$ spans an irreducible representation of S_n equivalent to V^λ or $\mathcal{S}^\lambda$, for which the set

$$\{\Delta(T) \mid T \text{ is a standard Young tableau with shape } \lambda\}$$

is a basis. It is an interesting exercise to show that applying the assignment of Eq. 10.40 from Sect. 10.4 to a standard polytabloid $\mathbf{e}_T$ gives the Specht polynomial $\Delta(T)$. See [HLV].

Decomposition of Young Permutation Modules

12

We have seen several examples of how Young permutation modules, realized as polynomial spaces, decompose into irreducible representations of the symmetric group. For example, as representations of S_3 we have

$$\mathcal{P}_{(1,0,0)} \cong \mathbb{C}[\mathcal{T}^{(2,1,0)}] \cong M^{(2,1,0)} \cong \mathcal{I} \oplus \mathcal{W} \cong V^{(3,0,0)} \oplus V^{(2,1,0)},$$

and

$$\mathcal{P}_{(2,1,0)} \cong \mathbb{C}[\mathcal{T}^{(1,1,1)}] \cong M^{(1,1,1)} \cong \mathcal{I} \oplus 2\,\mathcal{W} \oplus \mathcal{A} \cong V^{(3,0,0)} \oplus 2\,V^{(2,1,0)} \oplus V^{(1,1,1)}.$$

In this chapter we start by demonstrating a general method for determining the multiplicity of the irreducible components in Young permutation modules known as *Young's rule*. This involves extending some of the notions of Young tableaux, eventually constructing a basis for[1] $\mathrm{Hom}_{\mathbb{C}[S_n]}(\mathcal{S}^\lambda, \mathbb{C}[\mathcal{T}^\mu]) \cong \mathrm{Hom}_{\mathbb{C}[S_n]}(\mathcal{S}^\lambda, M^\mu)$, who's dimension gives the multiplicity of $\mathcal{S}^\lambda$ in $\mathbb{C}[\mathcal{T}^\mu]$ by Proposition 5.16. The discussion here is a mash-up (with a few modifications) of [Sa], Sections 2.8 and 2.9, [JK], Section 8.1, and [J], Chapter 13.

We then obtain an explicit basis for each of the irreducible components in polytabloid spaces and in polynomial spaces.

12.1 Generalized and Semistandard Young Tableaux

In this section we construct a new vector space that carries a representation of S_n, and show that it is equivalent to a polytabloid space $\mathbb{C}[\mathcal{T}^\mu]$.

---

[1] A space equivalent to.

© The Author(s), under exclusive license to Springer Nature Switzerland AG 2022
R. M. Howe, *An Invitation to Representation Theory*, SUMS Readings,
https://doi.org/10.1007/978-3-030-98025-2_12

Definition 12.1 Let $\lambda \vdash n$. A *generalized Young tableau* is a Young diagram with shape λ, where the boxes are filled with numbers from $\{1, 2, \ldots, n\}$, but where repetitions are allowed. A *semistandard Young tableau* is a generalized Young tableau where the entries are non-decreasing across the rows, and where the entries are strictly increasing down the columns.

For example, the tableaux

$$
\begin{array}{|c|c|c|}
\hline
3 & 1 & 1 \\
\hline
2 & 3 & 3 \\
\hline
\multicolumn{1}{|c|}{2} \\
\cline{1-1}
\multicolumn{1}{|c|}{5} \\
\cline{1-1}
\end{array}
\quad , \quad \text{and} \quad
\begin{array}{|c|c|c|}
\hline
1 & 1 & 2 \\
\hline
2 & 2 & 3 \\
\hline
\multicolumn{1}{|c|}{3} \\
\cline{1-1}
\multicolumn{1}{|c|}{4} \\
\cline{1-1}
\end{array}
$$

are, respectively, generalized and semistandard Young tableau with shape $(3, 3, 1, 1)$.

There are extra-special ways to fill semistandard Young tableaux.

Definition 12.2 Let $\lambda \vdash n$, and let $\mu = (\mu_1, \mu_2, \ldots, \mu_k) \models n$. Then a semistandard Young tableau with shape λ has *content* μ if there are μ_1 ones, μ_2 twos, and so on.

For example, the above tableaux with shape $(3,3,1,1)$ have content $(2, 2, 3, 0, 1)$ and $(2, 3, 2, 1)$ respectively.

Of special interest will be the case where both λ and μ are partitions of n.

Definition 12.3 Let λ and μ be partitions of n. The *Kotska numbers*, $K_{\lambda\mu}$, are defined as the number of semistandard Young tableaux with shape λ and content μ.

For example, $K_{(3.3.1.1)(3.2.1.1.1)} = 3$ since

$$
\begin{array}{|c|c|c|}
\hline
1 & 1 & 1 \\
\hline
2 & 2 & 3 \\
\hline
\multicolumn{1}{|c|}{4} \\
\cline{1-1}
\multicolumn{1}{|c|}{5} \\
\cline{1-1}
\end{array}
\quad , \quad
\begin{array}{|c|c|c|}
\hline
1 & 1 & 1 \\
\hline
2 & 2 & 4 \\
\hline
\multicolumn{1}{|c|}{3} \\
\cline{1-1}
\multicolumn{1}{|c|}{5} \\
\cline{1-1}
\end{array}
\quad , \quad \text{and} \quad
\begin{array}{|c|c|c|}
\hline
1 & 1 & 1 \\
\hline
2 & 2 & 5 \\
\hline
\multicolumn{1}{|c|}{3} \\
\cline{1-1}
\multicolumn{1}{|c|}{4} \\
\cline{1-1}
\end{array}
$$

are the only semistandard ways to fill the diagram $(3, 3, 1, 1)$ with 3 ones, 2 twos, 1 three, 1 four, and 1 five.

Exercise 12.4 Determine the value of $K_{\lambda\mu}$ for assorted values of n, λ, and μ.

Exercise 12.5 Let λ and μ be partitions of n. Show that $K_{(n)\mu} = 1$, that $K_{\lambda\lambda} = 1$, and that $K_{\lambda\mu} = 0$ whenever $\mu > \lambda$.

Remark 12.6 While there are other ways to characterize the Kotska numbers, there is no known explicit formula or algorithm for determining their value, and it is generally accepted that none exists.

12.2   The Space $\mathbb{C}[\mathcal{T}_{\lambda\mu}]$ and Its Equivalence to $\mathbb{C}[\mathcal{T}^{\mu}]$

Next we construct a vector space in a familiar way; let λ and μ be partitions of n, and let $\mathcal{T}_{\lambda\mu}$ denote the set of generalized Young tableau with shape λ and content μ. We declare these to be basis vectors, and then take all linear combinations over $\mathbb{C}$. We denote this space by $\mathbb{C}[\mathcal{T}_{\lambda\mu}]$. The goal in this section is to prove the following proposition.

Proposition 12.7 *The space $\mathbb{C}[\mathcal{T}^{\mu}]$ with the action described in Sect. 10.4, and the space $\mathbb{C}[\mathcal{T}_{\lambda\mu}]$ with the action described below, are equivalent representations of S_n.*

Proof For clarity, in this chapter we will indicate a fixed Young tableau, that is, a tableau with content (1^n), by a lowercase letter such as t or s, and use its entries to index the positions of the boxes. For example, given a fixed tableau

$$t = \begin{array}{|c|c|c|} \hline 1 & 3 & 5 \\ \hline 2 & 4 \\ \cline{1-2} \end{array},$$

we label the $(1, 2)$ entry with a 3, and the $(2, 2)$ entry with a 4. We then create a generalized Young tableau T by placing an entry $T(i)$ in the box labeled by i in t. For these first few examples we will write the *entry*, $T(i)$, in large type, and the *index*, i, in small type to the lower-right. For example, with t as above, and for

$$T = \begin{array}{|c|c|c|} \hline 2_{\,1} & 2_{\,3} & 1_{\,5} \\ \hline 1_{\,2} & 3_{\,4} \\ \cline{1-2} \end{array},$$

we have $T(1) = 2$, $T(2) = 1$, $T(3) = 2$, $T(4) = 3$, and $T(5) = 1$.

Now let μ and λ be partitions of n, and let $\{S\}$ be a Young tabloid with shape μ. We construct a tableau $T \in \mathcal{T}_{\lambda\mu}$ by defining

$T(i) :=$ the number of the row in which the index i (from t) appears in $\{S\}$.

For example, if $\mu = (2, 2, 1)$, and if

$$\{S\} = \begin{array}{|cc|} \hline 1 & 4 \\ \hline 3 & 2 \\ \hline 5 \\ \hline \end{array} \, ,$$

then

$$T = \begin{array}{|c|c|c|} \hline 1\,_1 & 2\,_3 & 3\,_5 \\ \hline 2\,_2 & 1\,_4 \\ \hline \end{array}$$

since the entries 1 and 4 both appear in the first row of $\{S\}$, the entries 2 and 3 both appear in the second row of $\{S\}$, and the entry 5 appears in the third row of $\{S\}$.

Exercise* 12.8 Work out a few examples, and convince yourself that:

(1) The map ϕ is well defined on tabloids.
(2) The shape of $\{S\}$ becomes the content of T.
(3) The map $\phi \colon \mathcal{T}^\mu \to \mathcal{T}_{\lambda\mu}$, given by $\{S\} \mapsto T$ as above, is a bijection between the basis vectors of the respective vector spaces. Be sure and write out the inverse mapping since we will use it later.
(4) Extending ϕ by linearity, conclude that $\mathbb{C}[\mathcal{T}^\mu]$ and $\mathbb{C}[\mathcal{T}_{\lambda\mu}]$ are isomorphic as vector spaces.

The question now becomes: how do we define an action of S_n on $\mathbb{C}[\mathcal{T}_{\lambda\mu}]$ in such a way that ϕ is an S_n-map? Since, for $\phi(\{S\}) = T$, our goal is to have

$$\phi(\sigma.\{S\}) = \sigma.[\phi(\{S\})] = \sigma.T,$$

we require that

$$\begin{aligned}
[\sigma.T](i) &= \text{the row number of } i \text{ in } \sigma.\{S\} \\
&= \text{the row number of } \sigma^{-1}(i) \text{ in } \{S\} \\
&= T(\sigma^{-1}i).
\end{aligned}$$

Keep in mind that $\sigma \in S_n$ acts on the entries of T by place permutation, that is, via the "index tableau" t (compare with Exercise 1.97). For example, if

$$T = \begin{array}{|c|c|c|} \hline 1\,_1 & 2\,_3 & 3\,_5 \\ \hline 2\,_2 & 1\,_4 \\ \hline \end{array} \, , \quad \text{then } (1, 3, 5).T = \begin{array}{|c|c|c|} \hline 3\,_5 & 1\,_1 & 2\,_3 \\ \hline 2\,_2 & 1\,_4 \\ \hline \end{array} \, .$$

Said another way,

$$\phi\left((1,3,5).\frac{\begin{array}{|c|c|}\hline 1 & 4 \\\hline 3 & 2 \\\hline 5 \\\cline{1-1}\end{array}}{}\right) = \phi\left(\begin{array}{|c|c|}\hline 3 & 4 \\\hline 5 & 2 \\\hline 1 \\\cline{1-1}\end{array}\right) = \begin{array}{|c|c|c|}\hline 3 & 1 & 2 \\\hline 2 & 1 \\\cline{1-2}\end{array}.$$

It follows that ϕ is an S_n-map, which completes the proof. $\square$

Exercise* 12.9 Here is a more enlightened way to obtain some of the results of this section. First check that S_n acts transitively on $\mathcal{T}_{\lambda\mu}$, and that the stabilizer (Exercise 1.62) of any $T \in \mathcal{T}_{\lambda\mu}$ is a Young subgroup of S_n that is isomorphic to S_μ. Conclude that the action of S_n on $\mathbb{C}[\mathcal{T}_{\lambda\mu}]$ is equivalent to the induced representation M^μ, and thus $\mathbb{C}[\mathcal{T}_{\lambda\mu}] \cong M^\mu \cong \mathbb{C}[\mathcal{T}^\mu]$ are equivalent as representations of S_n. A collateral result here is that the definition of M^μ does not require that $\mu_1 \geq \mu_2 \geq \cdots \geq 0$.

We will, however, have reason to make explicit use of the map ϕ from Exercise 12.8.

12.3   The Space $\text{Hom}_{\mathbb{C}[S_n]}(\mathcal{S}^\lambda, \mathbb{C}[\mathcal{T}_{\lambda\mu}])$

In this section, we construct S_n-intertwining maps from $\mathbb{C}[\mathcal{T}^\lambda]$ to $\mathbb{C}[\mathcal{T}_{\lambda\mu}]$ using generalized Young tableaux. In a later section we will show that those maps corresponding to semistandard Young tableaux constitute a basis for

$$\text{Hom}_{\mathbb{C}[S_n]}(\mathcal{S}^\lambda, \mathbb{C}[\mathcal{T}_{\lambda\mu}]) \cong \text{Hom}_{\mathbb{C}[S_n]}(\mathcal{S}^\lambda, \mathbb{C}[\mathcal{T}^\mu]) \cong \text{Hom}_{\mathbb{C}[S_n]}(\mathcal{S}^\lambda, M^\mu),$$

and the dimension of the space $\text{Hom}_{\mathbb{C}[S_n]}(\mathcal{S}^\lambda, M^\mu)$ gives the multiplicity of $\mathcal{S}^\lambda$ in M^μ (Proposition 5.16).

For $T \in \mathcal{T}_{\lambda,\mu}$ and $\{t\} \in \mathcal{T}^\lambda$, define

$$\theta_T : \mathbb{C}[\mathcal{T}^\lambda] \to \mathbb{C}[\mathcal{T}_{\lambda\mu}] \cong \mathbb{C}[\mathcal{T}^\mu], \quad \text{by } \theta_T(\{t\}) := \sum_{S \in \{T\}} S$$

on each basis vector $\{t\} \in \mathcal{T}^\lambda$, and then extend by linearity to all of $\mathbb{C}[\mathcal{T}^\lambda]$. Also, since for each $\{s\}$ and $\{t\}$ in $\mathcal{T}^\lambda$ we have $\{s\} = \{\sigma.t\}$ for some $\sigma \in S_n$, we can "extend θ_T by cyclicity." That is, we want the map θ_T to respect the group action and linearity:

$$\theta_T(\sigma.\{t\}) := \sigma.\theta_T(\{t\}) = \sigma. \sum_{S \in \{T\}} S = \sum_{S \in \{T\}} \sigma.S.$$

Example 12.10 Using the index tableau

$$t = \begin{array}{|c|c|c|} \hline 1 & 3 & 5 \\ \hline 2 & 4 \\ \cline{1-2} \end{array}, \quad \text{and for} \quad T = \begin{array}{|c|c|c|} \hline 2 & 1 & 1 \\ \hline 3 & 2 \\ \cline{1-2} \end{array} \in \mathbb{C}[\mathcal{T}_{\lambda\mu}],$$

we have

$$\theta_T(\{t\}) = \begin{array}{|c|c|c|} \hline 2 & 1 & 1 \\ \hline 3 & 2 \\ \cline{1-2} \end{array} + \begin{array}{|c|c|c|} \hline 1 & 2 & 1 \\ \hline 3 & 2 \\ \cline{1-2} \end{array} + \begin{array}{|c|c|c|} \hline 1 & 1 & 2 \\ \hline 3 & 2 \\ \cline{1-2} \end{array} + \begin{array}{|c|c|c|} \hline 2 & 1 & 1 \\ \hline 2 & 3 \\ \cline{1-2} \end{array} + \begin{array}{|c|c|c|} \hline 1 & 2 & 1 \\ \hline 2 & 3 \\ \cline{1-2} \end{array}$$

$$+ \begin{array}{|c|c|c|} \hline 1 & 1 & 2 \\ \hline 2 & 3 \\ \cline{1-2} \end{array},$$

and

$$\theta_T((1,5,4).\{t\}) := (1,5,4).\theta_T(\{t\})$$

$$= \begin{array}{|c|c|c|} \hline 2 & 1 & 2 \\ \hline 3 & 1 \\ \cline{1-2} \end{array} + \begin{array}{|c|c|c|} \hline 2 & 2 & 1 \\ \hline 3 & 1 \\ \cline{1-2} \end{array} + \begin{array}{|c|c|c|} \hline 2 & 1 & 1 \\ \hline 3 & 2 \\ \cline{1-2} \end{array} + \begin{array}{|c|c|c|} \hline 3 & 1 & 2 \\ \hline 2 & 1 \\ \cline{1-2} \end{array} + \begin{array}{|c|c|c|} \hline 3 & 2 & 1 \\ \hline 2 & 1 \\ \cline{1-2} \end{array}$$

$$+ \begin{array}{|c|c|c|} \hline 3 & 1 & 1 \\ \hline 2 & 2 \\ \cline{1-2} \end{array}.$$

Exercise* 12.11 For

$$t_2 = \begin{array}{|c|c|c|} \hline 1 & 2 & 3 \\ \hline 4 & 5 \\ \cline{1-2} \end{array}, \quad \text{and for} \quad T = \begin{array}{|c|c|c|} \hline 2 & 1 & 1 \\ \hline 3 & 2 \\ \cline{1-2} \end{array} \in \mathbb{C}[\mathcal{T}_{\lambda\mu}],$$

compute $\theta_T(\{t_2\})$ and $(1,5,4).\theta_T(\{t_2\})$. Compare the results for $\theta_T(\{t\})$ and $(1,5,4).\theta_T(\{t\})$ from Example 12.10. Check that, for any $T \in \mathbb{C}[\mathcal{T}_{\lambda\mu}]$, the value of $\theta_T(\{t\})$ is independent of the choice of $\{t\}$. Where does the choice of $\{t\}$ make a difference?

Exercise* 12.12 For

$$t_2' = \begin{array}{|c|c|c|} \hline 2 & 1 & 3 \\ \hline 4 & 5 \\ \cline{1-2} \end{array}, \quad \text{and for} \quad T = \begin{array}{|c|c|c|} \hline 2 & 1 & 1 \\ \hline 3 & 2 \\ \cline{1-2} \end{array} \in \mathcal{T}_{\lambda\mu},$$

compute $\theta_T(\{t_2'\})$ and $(1,5,4).\theta_T(\{t_2'\})$

Compare the results with Exercise 12.11. In particular, note that $\{t_2\} = \{t_2'\}$, and that $\theta_T(\{t_2\}) = \theta_T(\{t_2'\})$, but that $(1,5,4).\theta_T(\{t_2\}) \neq (1,5,4).\theta_T(\{t_2'\})$. Does this mean that action of S_n on θ_T is not well defined on equivalence classes in $\mathcal{T}^\lambda$?

Because we are ultimately interested in $\text{Hom}_{\mathbb{C}[S_n]}(\mathcal{S}^\lambda, \mathbb{C}[\mathcal{T}_{\lambda\mu}])$, we define $\overline{\theta}_T$ to be the restriction of θ_T to $\mathcal{S}^\lambda$, and consider the action of $\overline{\theta}_T$ on a standard polytabloid basis element $\mathbf{e}_t$ of $\mathcal{S}^\lambda$;

$$\overline{\theta}_T(\mathbf{e}_t) = \overline{\theta}_T\left(\mathcal{V}_t.\{t\}\right) = \mathcal{V}_t.\overline{\theta}_T(\{t\}) = \mathcal{V}_t.\left(\sum_{S \in \{T\}} S\right) = \sum_{S \in \{T\}} \mathcal{V}_t.S.$$

This righ-most expression could be zero, which would force $\overline{\theta}_T$ to be the zero-map since the $\mathbf{e}_t$'s cyclically generate $\mathcal{S}^\lambda$. We wish to eliminate those $T \in \mathcal{T}_{\lambda\mu}$ for which $\overline{\theta}_T$ could be identically zero. The next exercise illustrates the proposition that follows.

Exercise 12.13 Let

$$t = \begin{array}{|c|c|c|} \hline 1 & 4 & 5 \\ \hline 2 \\ \cline{1-1} 3 \\ \cline{1-1} \end{array}, \quad \text{and let} \quad S = \begin{array}{|c|c|c|} \hline 1 & 3 & 4 \\ \hline 1 \\ \cline{1-1} 2 \\ \cline{1-1} \end{array}.$$

Compute $\mathcal{V}_t.S$.

Proposition 12.14 *Let t be a fixed tableau with shape λ, and let $T \in \mathcal{T}_{\lambda\mu}$. Then $\mathcal{V}_t.S = 0$ if and only if S has two equal entries in the same column.*

Proof If $\mathcal{V}_t.S = 0$, then

$$\mathcal{V}_t.S = S + \sum_{\substack{\sigma \in V_t, \\ \sigma \neq (1)}} \text{sgn}(\sigma)\sigma.S = 0,$$

so we must (at least) have some $\sigma \in V_t$ such that $S = \sigma.S$ with $\text{sgn}(\sigma) = -1$. Since the elements in V_t permute the entries only along the columns of S, two of the entries in S must be equal and in the same column.

Now suppose that $i \neq j$ are in the same column of t, so that the entries $S(i) = S(j)$ are in the same column of S. Then we must have $[(1) - (i, j)].S = 0$. But by Exercise 11.49, $[(1) - (i, j)]$ is a factor of $\mathcal{V}_t$, and therefore $\mathcal{V}_t.S = 0$.   □

Corollary 12.15 *If T is a semistandard Young tableau, then $\overline{\theta}_T$ is not identically zero.*

With this corollary in mind, we designate an important subset of $\mathcal{T}_{\lambda\mu}$ by

$$\mathcal{T}_{\lambda\mu}^0 := \{T \in \mathcal{T}_{\lambda\mu} \mid T \text{ is semistandard}\}.$$

The following exercise produces examples that we will use later.

Exercise* 12.16 Let

$$t = \begin{array}{|c|c|c|}\hline 1 & 3 & 4 \\\hline 2 \\\cline{1-1}\end{array}, \quad T_1 = \begin{array}{|c|c|c|}\hline 1 & 1 & 3 \\\hline 2 \\\cline{1-1}\end{array}, \quad \text{and} \quad T_2 = \begin{array}{|c|c|c|}\hline 1 & 1 & 2 \\\hline 3 \\\cline{1-1}\end{array}.$$

Write out $\theta_{T_1}(\{t\})$, $\theta_{T_2}(\{t\})$, $\theta_{T_1}(\mathbf{e}_t)$, and $\theta_{T_2}(\mathbf{e}_t)$. You should also try some of your own examples.

12.4   Column Equivalence and Ordering

Analogous to row-equivalent tableaux (*i.e.*, tabloids), we declare two tableaux in $\mathcal{T}_{\lambda\mu}$ to be *column-equivalent* if they have the same entries in each column. We use vertical lines between the columns to indicate such an equivalence class, and let $[T]$ denote the column-equivalence class of T. For example,

$$T = \begin{array}{|c|c|c|c|}\hline 1 & 1 & 1 & 3 \\\hline 2 & 2 & 3 \\\cline{1-3} 4 \\\cline{1-1}\end{array} \sim \begin{array}{|c|c|c|c|}\hline 4 & 1 & 3 & 3 \\\hline 2 & 2 & 1 \\\cline{1-3} 1 \\\cline{1-1}\end{array} \sim \cdots \in \begin{array}{|c|c|c|c|} 1 & 1 & 1 & 3 \\ 2 & 2 & 3 \\ 4 \end{array} = [T].$$

Since "column-equivalence" makes sense only if the tableaux have the same shape and content, we can say more explicitly (referencing our fixed index-tableau t),

$$[T] := \{S \in \mathcal{T}_{\lambda\mu} \mid S = v.T \text{ for some } v \in V_t\}.$$

Define an order on this set of equivalence classes by declaring that $[T_1] > [T_2]$ if the smallest entry in $[T_1]$ appearing in a different column than in $[T_2]$ appears in an earlier (leftward) column of $[T_1]$ than in $[T_2]$. For example,

$$\begin{array}{|c|c|c|c|} 1 & 1 & 1 & 4 \\ 2 & 2 & 3 \\ 3 \end{array} \;>\; \begin{array}{|c|c|c|c|} 1 & 1 & 1 & 3 \\ 2 & 2 & 3 \\ 4 \end{array} \;>\; \begin{array}{|c|c|c|c|} 1 & 1 & 1 & 2 \\ 2 & 3 & 3 \\ 4 \end{array}.$$

Compare this with the order on tabloids given in Definition 11.24.

After a little thought, the following results should be clear:

- Two tableaux without the same shape and content may not be comparable. For example,

$$
\begin{array}{|c|c|c|c|}
\hline
1 & 1 & 1 & 2 \\
\hline
2 & 3 & 3 \\
\cline{1-3}
4 \\
\cline{1-1}
\end{array}
\quad \text{and} \quad
\begin{array}{|c|c|c|c|}
\hline
1 & 1 & 1 & 2 \\
\hline
2 & 2 & 3 \\
\cline{1-3}
4 \\
\cline{1-1}
\end{array}
$$

are not comparable.
- There is at most one semistandard Young tableau in each column-equivalence class, and there may be none. For example, the column-equivalence class

$$
\begin{array}{|c|c|c|c|}
\hline
1 & 1 & 3 & 2 \\
\hline
1 & 2 & 3 \\
\cline{1-3}
4 \\
\cline{1-1}
\end{array}
$$

contains no semistandard tableaux.
- This is a total order on the column-equivalence classes in $\mathcal{T}_{\lambda\mu}$, meaning that any two distinct equivalence classes are comparable. See Exercise 11.26.

Exercise 12.17 This exercise is trivial, but the result is crucial. Show that if T is semistandard, and if $T' \in \{T\}$, then $[T'] < [T]$ unless $T' = T$. It follows that if T is semistandard, then $[T]$ is the maximum column-equivalence class in the row-equivalence class $\{T\}$, and thus $[T]$ is the largest column-equivalence class appearing in $\theta_T(\{t\})$.

Remark 12.18 There are a several other definitions of "column-equivalence" and "partial order" in the literature that can be useful for obtaining further results. Any ordering that satisfies the results of Exercise 12.17 will work for our modest goals, and the order that we've defined here streamlines the proof.

Exercise 12.19 Let t be an index-tableau, that is, a standard Young tableau with shape λ, and let $\mathcal{V}_t$ be the column anti-symmetrizer of t. Check that, for a given tableau $T \in \mathcal{T}_{\lambda\mu}$ and every tableau T_i appearing in $\mathcal{V}_t.T$, we have $[T_i] = [T]$.

12.5   The Semistandard Basis for $\mathrm{Hom}_{\mathbb{C}[S_n]}(\mathcal{S}^\lambda, M^\mu)$

In this section we construct a basis for $\mathrm{Hom}_{\mathbb{C}[S_n]}(\mathcal{S}^\lambda, \mathbb{C}[\mathcal{T}_{\lambda\mu}]) \cong \mathrm{Hom}_{\mathbb{C}[S_n]}(\mathcal{S}^\lambda, M^\mu)$. The discussion is similar to the construction of the standard basis for Specht modules in Chap. 11. We will prove linear independence and span in separate propositions.

Proposition 12.20 *The set of semistandard homomorphisms, $\{\overline{\theta}_T \mid T \in \mathcal{T}^0_{\lambda\mu}\}$, is a linearly independent subset of* $\mathrm{Hom}_{\mathbb{C}[S_n]}(\mathcal{S}^\lambda, \mathbb{C}[\mathcal{T}_{\lambda\mu}])$.

It might be helpful to refer to the results of Exercise 12.16 while reading through the following proof.

Proof First recall that; from the definition of $\overline{\theta}_T$, the fact that T is semistandard, and Exercise 12.17, we have that $[T]$ is the maximum class appearing in $\overline{\theta}_T$ for each $T \in \mathcal{T}^0_{\lambda\mu}$.

Now consider the sum

$$\overline{\theta} := \sum c_T \overline{\theta}_T,$$

where each $T \in \mathcal{T}^0_{\lambda\mu}$, and where the c_T are scalars not all equal to zero. Our goal is to show that the map $\overline{\theta}$ is not identically zero. To this end, we choose T_1 such that the coefficient $c_{T_1} \neq 0$, but $c_T = 0$ whenever $[T] > [T_1]$. From this, and the remarks in the previous paragraph, it follows that

$$\overline{\theta} = \sum c_T \overline{\theta}_T = c_{T_1} T_1 + \{\text{a linear combination of tableaux } T_2 \text{ satisfying } [T_1] > [T_2]\}.$$

Applying $\overline{\theta}$ to a standard basis element $\mathbf{e}_t$ of $\mathcal{S}^\lambda$ yields

$$\overline{\theta}(\mathbf{e}_t) = \sum c_T \overline{\theta}_T(\mathbf{e}_t) = c_{T_1} \mathcal{V}_t . T_1$$

$$+ \{\text{a linear combination of expressions of the form } \mathcal{V}_t . T_3 \text{ where } [T_1] > [T_3]\}.$$

By Proposition 12.14, $\mathcal{V}_t . T_1 \neq 0$, and by Exercise 12.19, $\mathcal{V}_t$, the column stabilizer of t, preserves column-equivalence classes, so there is no way to "cancel" $\mathcal{V}_t . T_1$, and we have

$$\overline{\theta}(\mathbf{e}_t) = \sum c_T \overline{\theta}_T(\mathbf{e}_t) \neq 0.$$

Hence $\overline{\theta}$ is a non-zero element of $\mathrm{Hom}_{\mathbb{C}[S_n]}(\mathcal{S}^\lambda, \mathbb{C}[\mathcal{T}_{\lambda\mu}])$, as required, and we conclude that the set of semistandard homomorphisms is linearly independent.   □

Exercise 12.21 Verify that the semistandard homomorphisms in Exercise 12.16 are linearly independent.

Exercise* 12.22 Fill in the details in the above proof.

To show that the semistandard homomorphisms span $\mathrm{Hom}_{\mathbb{C}[S_n]}(\mathcal{S}^\lambda, \mathbb{C}[\mathcal{T}_{\lambda\mu}])$, we start with a somewhat technical lemma.

Lemma 12.23 *Let $\bar\theta$ be any non-zero element in $\mathrm{Hom}_{\mathbb{C}[S_n]}(\mathcal{S}^\lambda, \mathbb{C}[\mathcal{T}_{\lambda\mu}])$. Since $\bar\theta(\mathbf{e}_t) \in \mathbb{C}[\mathcal{T}_{\lambda\mu}]$, we can write*

$$\bar\theta(\mathbf{e}_t) = \sum c_T T \text{ for some scalars } c_T \text{ and tableaux } T \in \mathcal{T}_{\lambda\mu}.$$

Then:

(1) *The coefficients $c_{T^*} = 0$ for every tableau T^* having a repeated entry in some column.*

(2) *The coefficient $c_{T_1} \neq 0$ for some semistandard tableau T_1.*

Proof (1) Suppose that the indices $i \neq j$ are in the same column of t, suppose that the entries $T^*(i) = T^*(j)$ are in the same column of T^*, and consider the transposition $(i, j) \in V_t$. From Exercise 8.12 we have $v.\mathbf{e}_t = vV_t.\{t\} = \mathrm{sgn}(v)V_t.\{t\} = \mathrm{sgn}(v)\mathbf{e}_t$, and therefore $(i, j).\mathbf{e}_t = -\mathbf{e}_t$. Consequently,

$$\sum c_T (i, j).T = (i, j).\sum c_T T = (i, j).\theta(\mathbf{e}_t)$$
$$= \theta((i, j).\mathbf{e}_t) = \theta(-\mathbf{e}_t) = -\theta(\mathbf{e}_t) = -\sum c_T T.$$

Comparing coefficients in the expressions on each end, and using the fact that $(i, j).T^* = T^*$, we have $c_{T^*} = -c_{T^*}$. Thus $c_{T^*} = 0$ whenever T^* has a repeated entry in some column.

(2) Similar to part (1), for $v \in V_t$ we have $v.\mathbf{e}_t = \mathrm{sgn}(v)\mathbf{e}_t$, so

$$\sum c_T T = \sum c_T (\mathrm{sgn}\, v)v.T,$$

and therefore the $c_{T_1} = \pm c_{T_2}$ whenever T_1 and T_2 are column-equivalent.

Since $\bar\theta \neq 0$, we may choose a tableau T_1 such that $c_{T_1} \neq 0$ but $c_T = 0$ whenever $[T] > [T_1]$. By part (1) and the preceding paragraph, we may assume that the entries in T_1 are strictly increasing down the columns.

With the goal of obtaining a contradiction, suppose that T_1 is not semistandard, *i.e.*, that T_1 has a row-descent in some row q. More explicitly, some adjacent pair of columns in T_1, say the jth and $(j + 1)$st, have entries $\alpha_1 < \alpha_2 < \cdots < \alpha_r$ in column j, and entries and $\beta_1 < \beta_2 < \cdots < \beta_s$ in column $j + 1$, with $\alpha_q > \beta_q$ for some q. Of course, there is no need to make any claim regarding the entries of the index-tableau, t. The situation for the jth and

$(j + 1)$st columns of the tableaux t and T_1 looks like this:

$$
\begin{array}{cc@{\qquad\qquad\qquad}cc}
a_1 & b_1 & \alpha_1 & \beta_1 \\
 & & \wedge & \wedge \\
a_2 & b_2 & \alpha_2 & \beta_2 \\
 & & \wedge & \wedge \\
\cdot & \cdot & \cdot & \cdot \\
\cdot & \cdot & \cdot & \cdot \\
\cdot & \cdot & \cdot & \cdot \\
 & & \wedge & \wedge \\
t:\ a_q & b_q & T_1:\ \alpha_q > \beta_q \\
 & & \wedge & \wedge \\
\cdot & \cdot & \cdot & \cdot \\
\cdot & \cdot & \cdot & \cdot \\
\cdot & & \cdot & \wedge \\
\cdot & b_s & \cdot & \beta_s \\
\cdot & & \wedge \\
a_r & & \alpha_r
\end{array}
$$

We then form the Garnir element $g_{A,B}$ as in Sect. 11.4 and apply it to $\overline{\theta}(\mathbf{e}_t)$. It is important to remember that for each tableau T, $g_{A,B}$ acts on T by PLACE permutations via the index tableau t, so we use the sets $A = \{a_q, \ldots, a_r\}$ and $B = \{b_1, \ldots, b_q\}$ from t to form $g_{A,B}$.

It is useful here to pause for a simple exercise.

Exercise* 12.24 Let

$$
t = \begin{array}{|c|c|c|} \hline 1 & 3 & 5 \\ \hline 2 & 4 \\ \cline{1-2} \end{array}
\quad \text{and} \quad
T_1 = \begin{array}{|c|c|c|} \hline 1 & 2 & 1 \\ \hline 2 & 3 \\ \cline{1-2} \end{array}.
$$

Form the Garnir element $g_{A,B}$ for the tableau T_1 and write out $g_{A,B}.T_1$. You should also try a more complicated example.

Continuing our proof, by Proposition 11.50

$$
g_{A,B}.\overline{\theta}(\mathbf{e}_t) = \overline{\theta}(g_{A,B}.\mathbf{e}_t) = \overline{\theta}(0) = 0.
$$

Referring to the statement of the lemma, this implies that

$$
g_{A,B}.\overline{\theta}(\mathbf{e}_t) = g_{A,B}. \sum c_T T = \sum c_T g_{A,B}.T = 0.
$$

But since for any tableau T, all of the tableaux appearing in $g_{A,B}.T$ have coefficient ± 1, and since $\sum c_T g_{A,B}.T = 0$, there must be a tableau $T' \neq T_1$ with coefficient $c_{T'} \neq 0$, along with a permutation π in $g_{A,B}$, so that $\pi.T' = T_1$. It follows that the entries of T' must agree with the entries of T_1 in each position except those positions indexed by A and B (from the tableau t).

Now $\beta_1 < \cdots < \beta_q < \alpha_q < \cdots < \alpha_r$, and the permutation π switches some β entries with some α entries, so there must be a minimum entry β_i that appears in column in j of T' and that also appears in column $j+1$ of $\pi T' = T_1$. Therefore $[T'] > [T_1]$, and this contradicts our choice of T_1. $\square$

Exercise* 12.25 Let

$$t = \begin{array}{|c|c|c|} \hline 1 & 3 & 5 \\ \hline 2 & 4 \\ \cline{1-2} \end{array} \quad \text{and} \quad T_1 = \begin{array}{|c|c|c|} \hline 1 & 2 & 1 \\ \hline 2 & 3 \\ \cline{1-2} \end{array}.$$

Form the Garnir element $g_{A,B}$ for the tableau T_1. Find an element π appearing in $g_{A,B}$ and a tableau T such that $\pi.T = T_1$. You should also try a more complicated example.

Proposition 12.26 *The set of semistandard homomorphisms $\{\bar\theta_T \mid T \in \mathcal{T}^0_{\lambda\mu}\}$ spans* $\text{Hom}_{\mathbb{C}[S_n]}(\mathcal{S}^\lambda, \mathbb{C}[\mathcal{T}_{\lambda\mu}]) \cong \text{Hom}_{\mathbb{C}[S_n]}(\mathcal{S}^\lambda, M^\mu).$

Proof Let $\bar\theta \in \text{Hom}_{\mathbb{C}[S_n]}(\mathcal{S}^\lambda, \mathbb{C}[\mathcal{T}_{\lambda\mu}])$ be non-zero. By Lemma 12.23, $\bar\theta(\mathbf{e}_t) = \sum c_T T$, where $c_{T_1} \neq 0$ for some semistandard $T_1 \in \mathcal{T}^0_{\lambda\mu}$, and, since there must be a largest such tableau, we may assume that $c_T = 0$ whenever $T \in \mathcal{T}^0_{\lambda\mu}$ and $[T] > [T_1]$. Now, by Exercise 12.17,

$$\bar\theta_{T_1} = T_1 + \{\text{a sum of tableaux of } T_2 \text{ where } [T_1] > [T_2]\},$$

and it follows that $(\bar\theta - c_{T_1}\bar\theta_{T_1})(\mathbf{e}_t)$ is a linear combination of tableaux T_2 with $[T_2] < [T_1]$. By induction, the homomorphism $(\bar\theta - c_{T_1}\bar\theta_{T_1})$ is a linear combination of semistandard homomorphisms, and therefore so is $\bar\theta$. $\square$

Corollary 12.27 *Propositions 12.20 and 12.26 along with Proposition 5.16 yield*

$$\text{Dim } \text{Hom}_{\mathbb{C}[S_n]}(\mathcal{S}^\lambda, M^\mu) = K_{\lambda\mu} = \text{the multiplicity of } \mathcal{S}^\lambda \text{ in } M^\mu.$$

12.6   Young's Rule

There are several versions of Young's rule that arise in different contexts related
to the symmetric group, and this is the version that applies to our situation.[2] It is
important enough to state as a theorem, but it is an immediate corollary to the results
of Sect. 12.5.

Theorem 12.28 (Young's Rule) *Let $\mu \vdash n$, and let M^{μ} be the Young permutation
module induced from the trivial representation of the Young subgroup S_{μ}. For $\lambda \vdash n$,
let $K_{\lambda\mu}$ be as in Definition 12.3, and let $S^{\lambda} \cong V^{\lambda}$ be an irreducible representation
of S_n labeled by λ. Then*

$$M^{\mu} \cong \bigoplus_{\lambda \vdash n} K_{\lambda\mu} S^{\lambda}.$$

Example 12.29 We illustrate this theorem with the familiar permutation modules
for S_3.

- For $M^{(3,0,0)}$;

$$\boxed{1}\boxed{1}\boxed{1}$$

is the only semistandard way to fill any tableau $\lambda \vdash 3$ with content $(3, 0, 0)$, so

$$M^{(3,0,0)} \cong S^{(3,0,0)}.$$

- For $M^{(2,1,0)}$;

$$\boxed{1}\boxed{1}\boxed{2} \quad \text{and} \quad \begin{array}{c}\boxed{1}\boxed{1}\\\boxed{2}\end{array}$$

are the only only semistandard ways to fill tableaux $\lambda \vdash 3$ with content $(2, 1, 0)$,
so

$$M^{(2,1,0)} \cong S^{(3,0,0)} \oplus S^{(2,1,0)}.$$

---

[2] There is also a "Young's Rule" in medicine.

- For $M^{(1,1,1)}$:

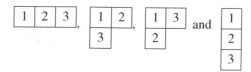

are the only semistandard ways to fill tableaux $\lambda \vdash 3$ with content $(1, 1, 1)$, so

$$M^{(1,1,1)} \cong S^{(3,0,0)} \oplus 2\, S^{(2,1,0)} \oplus S^{(1,1,1)}.$$

At the risk of stating the obvious, this last case reflects the fact that $M^{(1^n)}$ is equivalent to the left regular representation on $\mathbb{C}[S_n]$, and that $K_{\lambda(1^n)} = f^\lambda$.

Exercise* 12.30 Decompose the polynomial space $\mathcal{P}_4(x, y, z, w)$ into irreducible representations of S_4.

Exercise* 12.31 Use the map $\bar{\theta}_T : S^\lambda \to \mathbb{C}[\mathcal{T}_{\lambda\mu}]$, the map $\phi^{-1} : \mathbb{C}[\mathcal{T}_{\lambda\mu}] \to \mathbb{C}[\mathcal{T}^\mu]$ from Exercise 12.8, and the assignment $\{T\} \in \mathcal{T}^\mu$ with a monomial as in Eq. 10.40 to obtain an explicit basis for some of the irreducible components in $\mathcal{P}_4(x, y, z, w)$.

Exercise* 12.32 Apply the intertwining maps from Exercises 3.21–3.23 to your results from Exercise 12.31 to produce bases for irreducible representations in other polynomial spaces. Can you find other intertwining maps like these?

12.7 Hints and Additional Comments

Exercise 12.8 The map $\phi^{-1} : \mathbb{C}[\mathcal{T}_{\lambda\mu}] \to \mathbb{C}[\mathcal{T}^\mu]$ is given by "place the corresponding entry from the index tableau in this row of the tabloid." For

$$t = \begin{array}{|c|c|c|}\hline 1 & 3 & 5 \\\hline 2 & 4 \\\cline{1-2}\end{array} \quad \text{we have} \quad \phi^{-1}\left(\begin{array}{|c|c|c|}\hline 1 & 1 & 2 \\\hline 2 & 3 \\\cline{1-2}\end{array}\right) = \begin{array}{cc} 1 & 3 \\ \hline 2 & 5 \\ \hline 4 \\ \hline \end{array}$$

since the index entries 1 and 3 get placed in the first row, the entries 2 and 5 in the second row, and the entry 4 gets placed in the third row.

Exercise 12.9 Let t be a fixed index tableau, and choose a tableau $T \in \mathcal{T}_{\lambda\mu}$. For example,

$$t = \begin{array}{|c|c|c|} \hline 1 & 3 & 5 \\ \hline 2 & 4 \\ \cline{1-2} \end{array} \quad \text{and } T = \begin{array}{|c|c|c|} \hline 1 & 1 & 2 \\ \hline 2 & 3 \\ \cline{1-2} \end{array}.$$

Since S_n acts by place permutation, it is easy to see that any other tableau with the same shape and content can be obtained from T by such a permutation, and so the action of S_n is transitive on $\mathcal{T}_{\lambda\mu}$. In this example, the stabilizer of T is $S_\mu = S_{\{1,3\}} \times S_{\{2,5\}} \times S_{\{4\}} \cong S_{(2,2,1)}$.

It is routine to check that the map S_n/S_μ given by $\sigma.S_\mu \mapsto \sigma.T$ is well defined on cosets. Recall (Exercise 10.38) that the representation of S_n on S_n/S_μ is equivalent to the representation of S_n on $\mathbb{C}[T^\mu]$, so we have an equivalence between the coset representation of S_n on S_n/S_μ and the representation of S_n on $\mathbb{C}[\mathcal{T}_{\lambda\mu}]$.

Exercise 12.11

$$\theta_T(\{t_2\}) = \begin{array}{|c|c|c|} \hline 2 & 1 & 1 \\ \hline 3 & 2 \\ \cline{1-2} \end{array} + \begin{array}{|c|c|c|} \hline 1 & 2 & 1 \\ \hline 3 & 2 \\ \cline{1-2} \end{array} + \begin{array}{|c|c|c|} \hline 1 & 1 & 2 \\ \hline 3 & 2 \\ \cline{1-2} \end{array} + \begin{array}{|c|c|c|} \hline 2 & 1 & 1 \\ \hline 2 & 3 \\ \cline{1-2} \end{array} + \begin{array}{|c|c|c|} \hline 1 & 2 & 1 \\ \hline 2 & 3 \\ \cline{1-2} \end{array}$$

$$+ \begin{array}{|c|c|c|} \hline 1 & 1 & 2 \\ \hline 2 & 3 \\ \cline{1-2} \end{array}.$$

$$(1,5,4).\theta_T(\{t_2\}) = \begin{array}{|c|c|c|} \hline 3 & 1 & 1 \\ \hline 2 & 2 \\ \cline{1-2} \end{array} + \begin{array}{|c|c|c|} \hline 3 & 2 & 1 \\ \hline 2 & 1 \\ \cline{1-2} \end{array} + \begin{array}{|c|c|c|} \hline 3 & 1 & 2 \\ \hline 2 & 1 \\ \cline{1-2} \end{array} + \begin{array}{|c|c|c|} \hline 2 & 1 & 1 \\ \hline 3 & 2 \\ \cline{1-2} \end{array}$$

$$+ \begin{array}{|c|c|c|} \hline 2 & 2 & 1 \\ \hline 3 & 1 \\ \cline{1-2} \end{array} + \begin{array}{|c|c|c|} \hline 2 & 1 & 2 \\ \hline 3 & 1 \\ \cline{1-2} \end{array}.$$

Notice that $\theta_T(\{t_2\}) = \theta_T(\{t\})$, but that $(1,5,4).\theta_T(\{t_2\}) \neq (1,5,4).\theta_T(\{t\})$, which is to be expected since $\{t\} \neq \{t_2\}$.

Exercise 12.12

$$\theta_T(\{t_2'\}) = \begin{array}{|c|c|c|} \hline 2 & 1 & 1 \\ \hline 3 & 2 \\ \cline{1-2} \end{array} + \begin{array}{|c|c|c|} \hline 1 & 2 & 1 \\ \hline 3 & 2 \\ \cline{1-2} \end{array} + \begin{array}{|c|c|c|} \hline 1 & 1 & 2 \\ \hline 3 & 2 \\ \cline{1-2} \end{array} + \begin{array}{|c|c|c|} \hline 2 & 1 & 1 \\ \hline 2 & 3 \\ \cline{1-2} \end{array} + \begin{array}{|c|c|c|} \hline 1 & 2 & 1 \\ \hline 2 & 3 \\ \cline{1-2} \end{array}$$

$$+ \begin{array}{|c|c|c|} \hline 1 & 1 & 2 \\ \hline 2 & 3 \\ \cline{1-2} \end{array}.$$

$$(1,5,4).\theta_T(\{t_2'\}) = \boxed{\begin{array}{ccc}2&3&1\\2&1\end{array}} + \boxed{\begin{array}{ccc}1&3&1\\2&2\end{array}} + \boxed{\begin{array}{ccc}1&3&2\\2&1\end{array}} + \boxed{\begin{array}{ccc}2&2&1\\3&1\end{array}}$$

$$+ \boxed{\begin{array}{ccc}1&2&1\\3&2\end{array}} + \boxed{\begin{array}{ccc}1&2&2\\3&1\end{array}}.$$

Even though $\{t_2\} = \{t_2'\}$, $(1,5,4).\theta_T(\{t_2\})$ and $(1,5,4).\theta_T(\{t_2'\})$ are not even close! What's up?

The key is to next apply the map ϕ^{-1} from Exercise 12.8, which yields the desired equality. Note also that this equality does not occur in the previous case where $\{t\} \neq \{t_2\}$.

$$\phi^{-1}((1,5,4).\theta_T\{t_2\})$$

$$= \boxed{\begin{array}{cc}3&5\\2&4\\1\end{array}} + \boxed{\begin{array}{cc}2&3\\4&5\\1\end{array}} + \boxed{\begin{array}{cc}2&5\\3&4\\1\end{array}} + \boxed{\begin{array}{cc}3&5\\1&2\\4\end{array}} + \boxed{\begin{array}{cc}2&3\\1&5\\4\end{array}} + \boxed{\begin{array}{cc}2&5\\1&3\\4\end{array}}$$

$$= \phi^{-1}((1,5,4).\theta_T\{t_2'\}).$$

Exercise 12.16 Note that T_1 and T_2 are the only semistandard tableaux with the given shape and content, and that $[T_1] > [T_2]$.

$$\theta_{T_1}(\{t\}) = \sum_{S \in \{T_1\}} S = \boxed{\begin{array}{ccc}1&1&3\\2\end{array}} + \boxed{\begin{array}{ccc}1&3&1\\2\end{array}} + \boxed{\begin{array}{ccc}3&1&1\\2\end{array}}.$$

$$\theta_{T_2}(\{t\}) = \sum_{S \in \{T_2\}} S = \boxed{\begin{array}{ccc}1&1&2\\3\end{array}} + \boxed{\begin{array}{ccc}1&2&1\\3\end{array}} + \boxed{\begin{array}{ccc}2&1&1\\3\end{array}}.$$

$$\theta_{T_1}(\mathbf{e}_t) = \theta_{T_1}([(1) - (1,2)].\{t\}) := [(1) - (1,2)].\theta_{T_1}(\{t\})$$

$$= \boxed{\begin{array}{ccc}1&1&3\\2\end{array}} + \boxed{\begin{array}{ccc}1&3&1\\2\end{array}} + \boxed{\begin{array}{ccc}3&1&1\\2\end{array}}$$

$$- \boxed{\begin{array}{ccc}2&1&3\\1\end{array}} - \boxed{\begin{array}{ccc}2&3&1\\1\end{array}} - \boxed{\begin{array}{ccc}2&1&1\\3\end{array}}.$$

$$\theta_{T_2}(\mathbf{e}_t) = \theta_{T_2}([(1) - (1,2)].\{t\}) := [(1) - (1,2)].\theta_{T_2}(\{t\})$$

$$= \young(112,3) + \young(121,3) + \young(211,3)$$

$$- \young(312,1) - \young(321,1) - \young(311,2).$$

Exercise 12.22 Since we have a total order on column-equivalence classes, and $[T_1]$ is maximum, we have

$$\bar{\theta} = \sum c_T \bar{\theta}_T = c_{T_1} \bar{\theta}_{T_1} + \{\text{a linear combination of homomorphisms } \bar{\theta}_{T_2}, \text{ where } [T_2] < [T_1]\}.$$

Applying $\bar{\theta}$ to a standard basis element $\mathbf{e}_t$ in $\mathcal{S}^\lambda$;

$$\bar{\theta}(\mathbf{e}_t) = c_{T_1} \bar{\theta}_{T_1}(\mathbf{e}_t) + \{\text{a linear combination of expressions } \bar{\theta}_{T_2}(\mathbf{e}_t), \text{ where } [T_2] < [T_1]\}.$$

Substituting $\mathbf{e}_t = \mathcal{V}_t.\{t\}$ yields

$$\bar{\theta}(\mathbf{e}_t) = c_{T_1} \bar{\theta}_{T_1}(\mathcal{V}_t.\{t\}) + \{\text{a linear combination of expressions } \bar{\theta}_{T_2}(\mathcal{V}_t.\{t\}), \text{ where } [T_2] < [T_1]\}.$$

Since any $\bar{\theta}_T$ intertwines the action of $\mathcal{V}_t \in \mathbb{C}[S_n]$, we have

$$\bar{\theta}(\mathbf{e}_t) = c_{T_1} \mathcal{V}_t.\bar{\theta}_{T_1}(\{t\}) + \{\text{a linear combination of expressions } \mathcal{V}_t.\bar{\theta}_{T_2}(\{t\}), \text{ where } [T_2] < [T_1]\}.$$

The results of Exercise 12.17 imply that

$$\bar{\theta}(\mathbf{e}_t) = c_{T_1} \mathcal{V}_t.(T_1 + \{\text{a sum of tableaux } T_i \text{ satisfying } [T_i] < [T_1]\})$$
$$+\{\text{a linear combination of expressions } \mathcal{V}_t.T_j \text{ where } [T_j] < [T_1]\}$$
$$= c_{T_1} \mathcal{V}_t.T_1 + c_{T_1}(\{\text{a sum of expressions } \mathcal{V}_t.T_i \text{ satisfying } [T_i] < [T_1]\})$$
$$+\{\text{a linear combination of expressions } \mathcal{V}_t.T_j, \text{ where } [T_j] < [T_1]\}.$$

Exercise 12.24 In this example, $g_{A.B} = (1) + (4,5)(3,4) - (3,5)$, and hence

$$g_{A.B}.T_1 = \young(121,23) + \young(113,22) - \young(112,23).$$

Exercise 12.25 From the results of Exercise 12.24,

$$(3,5).\young(112,23) = \young(121,23) = T_1.$$

Note that if π is a permutation appearing in $g_{A,B}$, and if $\pi.T = T_1$, then $[T] > [T_1]$ since the smallest entry of $[T]$ moved by π is to the right.

Exercise 12.30

$$\mathcal{P}_4(x, y, z, w) = \mathcal{P}_{(4,0,0,0)} \oplus \mathcal{P}_{(3,1,0,0)} \oplus \mathcal{P}_{(2,2,0,0)} \oplus \mathcal{P}_{(2,1,1,0)} \oplus \mathcal{P}_{(1,1,1,1)}.$$

Using the results of Sect. 10.3 we have;

$$\mathcal{P}_{(4,0,0,0)} \cong M^{(3,1,0,0)},$$
$$\mathcal{P}_{(3,1,0,0)} \cong M^{(2,1,1,0)},$$
$$\mathcal{P}_{(2,2,0,0)} \cong M^{(2,2,0,0)},$$
$$\mathcal{P}_{(2,1,1,0)} \cong M^{(2,1,1,0)},$$
$$\text{and } \mathcal{P}_{(1,1,1,1)} \cong M^{(4,0,0,0)}.$$

Some of the computations below reproduce the results from Exercise 12.5. Working our way through the permutation modules in lexicographic order:

- For $M^{(4,0,0,0)}$;

$$\boxed{1}\,\boxed{1}\,\boxed{1}\,\boxed{1}$$

is the only semistandard way to fill the tableau $(4, 0, 0, 0)$ with content $(4, 0, 0, 0)$, and $K_{\lambda,(4,0,0,0)} = 0$ for all other $\lambda \vdash 4$. Hence

$$M^{(4,0,0,0)} \cong S^{(4,0,0,0)}.$$

- For $M^{(3,1,0,0)}$;

$$\boxed{1}\,\boxed{1}\,\boxed{1}\,\boxed{2}$$

is the only semistandard way to fill the tableau $(4, 0, 0, 0)$ with content $(3, 1, 0, 0)$, and

$$\begin{array}{ccc}\boxed{1}&\boxed{1}&\boxed{1}\\\boxed{2}\end{array}$$

is the only semistandard way to fill the tableau $(3, 1, 0, 0)$ with content $(3, 1, 0, 0)$. All other values of $K_{\lambda,(3,1,0,0)}$ for $\lambda < (3, 1, 0, 0)$ are zero, so

$$M^{(3,1,0,0)} \cong S^{(4,0,0,0)} \oplus S^{(3,1,0,0)}.$$

- For $M^{(2,2,0,0)}$ the various semistandard tableaux are;

$$\boxed{\begin{array}{|c|c|c|c|}\hline 1 & 1 & 2 & 2 \\\hline\end{array}}, \quad \begin{array}{|c|c|c|}\hline 1 & 1 & 2 \\\hline 2 \\\cline{1-1}\end{array}, \quad \text{and} \quad \begin{array}{|c|c|}\hline 1 & 1 \\\hline 2 & 2 \\\hline\end{array}.$$

Hence

$$M^{(2,2,0,0)} \cong \mathcal{S}^{(4,0,0,0)} \oplus \mathcal{S}^{(3,1,0,0)} \oplus \mathcal{S}^{(2,2,0,0)}.$$

- For $M^{(2,1,1,0)}$ the various semistandard tableaux are;

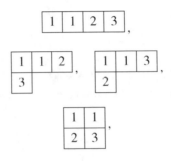

and

$$\begin{array}{|c|c|}\hline 1 & 1 \\\hline 2 \\\cline{1-1} 3 \\\cline{1-1}\end{array}.$$

Therefore,

$$M^{(2,1,1,0)} \cong \mathcal{S}^{(4,0,0,0)} \oplus 2\,\mathcal{S}^{(3,1,0,0)} \oplus \mathcal{S}^{(2,2,0,0)} \oplus \mathcal{S}^{(2,1,1,0)}.$$

Summarizing this whole mess yields;

$$\mathcal{P}_4(x,y,z,w) \cong M^{(4,0,0,0)} \oplus M^{(3,1,0,0)} \oplus M^{(2,2,0,0)} \oplus 2\,M^{(2,1,1,0)}$$
$$\cong 5\,\mathcal{S}^{(4,0,0,0)} \oplus 6\,\mathcal{S}^{(3,1,0,0)} \oplus 3\,\mathcal{S}^{(2,2,0,0)} \oplus 2\,\mathcal{S}^{(2,1,1,0)}.$$

Now consider the invariant subspace

$$\mathcal{P}_{(2,1,1,0)} = \langle\!\langle \mathbf{x^2yz}, \ldots, \mathbf{y^2zw} \rangle\!\rangle \cong M^{(2,1,1,0)}.$$

By Young's rule, this space contains two copies of the Specht module $\mathcal{S}^{(3,1,0,0)} \in \mathbb{C}[\mathcal{T}^{(3,1,0,0)}]$ labeled by the two semi-standard tableaux with shape $(3, 1, 0, 0)$ and content $(2, 1, 1, 0)$;

$$T_1 = \young(112,3) \quad \text{and} \quad T_2 = \young(113,2).$$

The Specht module $\mathcal{S}^{(3,1,0,0)}$ has as a basis the three standard polytabloids e_{t_i} where

$$t_1 = \young(134,2), \quad t_2 = \young(124,3) \quad \text{and} \quad t_3 = \young(123,4).$$

For a first step, we apply $\bar{\theta}_{T_1}$ to e_{t_1} to obtain a basis vector for the copy of $\mathcal{S}^{(3,1,0,0)}$ in $\mathbb{C}[\mathcal{T}_{(3,1)(2,1,1)}]$;

$$\bar{\theta}_{T_1}(e_{t_1}) = \sum_{S \in \{T_1\}} \mathcal{V}_{t_1}.S$$

$$= \young(112,3) + \young(121,3) + \young(211,3)$$

$$- \young(312,1) - \young(321,1) - \young(311,2}.$$

We next apply ϕ^{-1} from Exercise 12.8, which yields a basis vector for this copy of $\mathcal{S}^{(3,1,0,0)}$ in $\mathbb{C}[\mathcal{T}^{(2,1,1,0)}]$;

$$\young(13,4,2) + \young(14,3,2) + \young(34,1,2) - \young(23,4,1) - \young(24,3,1) - \young(34,2,1}.$$

Finally, using the entries as subscripts and applying the method from Example 10.41 with $x_1 = x$, $x_2 = y$, $x_3 = z$, and $x_4 = w$ yields a basis polynomial

$$xzw^2 + xz^2w + x^2zw - yzw^2 - yz^2w - y^2zw \in \mathcal{P}_{(2,1,1,0)},$$

where, for example, the first and last terms are obtained via

$$\young(xz,w,y)_{102} \quad \text{and} \quad \young(zw,y,x)_{120}.$$

Performing the same steps to $\mathbf{e}_{t_2}$ and $\mathbf{e}_{t_3}$ gives us a basis for the copy of $\mathcal{S}^{(3,1,0,0)}$ in $\mathcal{P}_{(2,1,1,0)}$ labeled by the semistandard tableau T_1.

Using $\overline{\theta}_{T_2}$ for the first step, we can obtain a basis for the other copy of $\mathcal{S}^{(3,1,0,0)}$ in $\mathcal{P}_{(2,1,1,0)}$. By mimicking this procedure we can completely decompose $\mathcal{P}_4$, obtaining a basis for each of the irreducible components.

Exercise 12.32 Expressions such as $x^2 y \partial_x \partial_z^2$, that act linearly on function spaces, are examples of *polynomial coefficient differential operators*. The collection of all linear combinations of these operators, denoted $\mathcal{PD}$, is often referred to as the *Weyl algebra*, after the mathematician and theoretical physicist Hermann Weyl.

The obvious question is, "How do we define an action of S_n on $\mathcal{PD}$?" First observe that, since $\partial_{x_i} x_j = \delta_{i,j}$, the set $\{\partial_{x_1}, \ldots, \partial_{x_n}\}$ is the basis dual to the basis $\{x_1, \ldots, x_n\}$ of $\mathcal{P}_1$. As with any dual basis, consistency requires that $\sigma . \partial_{x_i}(\sigma . x_j) = \partial_{x_i} x_j$, so it follows that $\sigma . \partial_{x_i} = \partial_{x_{\sigma(i)}}$, and we can extend this action in the obvious way, as we do with polynomials as in Sect. 2.1. For example,

$$(1, 2, 3) . x^2 y \partial_x \partial_z^2 = y^2 z \partial_y \partial_x^2.$$

It should be routine to check that $f \in \mathcal{PD}$ is an intertwining map exactly when $\sigma . f = f$. An important example is the *degree operator*: $f = \sum x_i \partial_{x_i}$.

13

This chapter demonstrates how the irreducible representations of S_n decompose when induced or restricted. The basic idea is that the Young symmetrizers for S_n (Sect. 8.3) can be restricted to S_{n-1} to produce irreducible representations of S_{n-1}. Counting dimensions then yields the decomposition of restricted representations, and applying Frobenius reciprocity yields the decomposition of induced representations.

13.1   The Hook Length Formula

Here is a remarkable formula for f^λ, the number of standard Young tableaux with shape λ, which is equal to the dimension of the irreducible S_n-module $\mathcal{S}^\lambda \cong V^\lambda$. There are a number of quite different proofs of this result, and the presentation here follows [F], Chapter 4, Exercise 11, and [K], Section 5.1.4. See also [GN].

Exercise* 13.1 Show that $f^\lambda = f^{\lambda'}$, where λ' is the partition conjugate to λ (Sect. 8.1).

We start with a definition that may seem somewhat strange at first.

Definition 13.2 Given a box (i, j) in a Young diagram λ, a *hook* $H_{i,j}$ is the box (i, j), along with the boxes in its row to the right and the boxes in its column below. The number of boxes in a hook, $h(i, j) := |H_{i,j}|$, is called the *hook length*. The hook $H_{2,3}$ with $h(2, 3) = 6$ is depicted in the example below, as are the hook lengths for some of the other boxes.

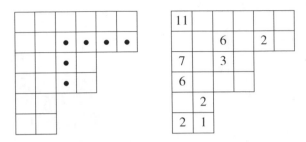

Exercise 13.3 Fill in some of the other hook lengths in the above tableau.

Proposition 13.4 (Hook Length Formula) *If λ is a Young diagram with n boxes, then*

Equation 13.5

$$f^{\lambda} = \frac{n!}{\prod h(i, j)}.$$

In other words, the number of standard Young tableau with shape $\lambda \vdash n$ equals n! divided by the product of the hook lengths from each box of λ.

Towards establishing Proposition 13.4, we say that if $\lambda \vdash n$ is a Young diagram, an *outer corner* of λ is a box that lies at the end of a row and that also lies at the bottom of a column. We denote by λ^{-} any Young diagram that can be obtained from λ by deleting such an outer corner, and it is easy to check that $\lambda^{-} \vdash n - 1$ is a legitimate Young diagram. Below we've bulleted these outer corners,

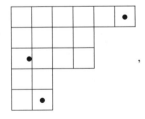

,

and we use the notation $\lambda^{-} \prec \lambda$ when λ^{-} can be obtained from λ by deleting an outer corner.

Since the box containing n in a standard Young tableau must be an outer corner, and since designating a standard tableau with n boxes is the same as designating one with $n - 1$ boxes and then deciding where to put the nth box, we have the following recursive formula.

Equation 13.6

$$f^\lambda = \sum_{\lambda^- \lessdot \lambda} f^{\lambda^-}.$$

Example 13.7 Consider a standard tableau with shape $(3, 2)$ and observe that there are only two possible choices of where to place the entry 5;

Since there are two standard tableaux with shape $(2, 2)$, and three standard tableaux with shape $(3, 1)$, there are $2 + 3 = 5$ standard tableau with shape $(3, 2)$, as we have previously observed.

Exercise 13.8 Use Exercise 13.1 and the reasoning from Example 13.7 to determine $f^{(3,2,1)}$. Check your result with the hook length formula.

We next establish a formula for the product of hook lengths. If $\lambda = (\lambda_1, \ldots, \lambda_k)$, define $\ell_i := \lambda_i + k - i$ for each $i = 1 \ldots k$.

Exercise 13.9 Check that the number ℓ_i is the hook length for the first box in the ith row of the Young diagram for λ, and that $\ell_k = \lambda_k$. Also check that if $\lambda_i = \lambda_{i+1}$, then $\ell_i - \ell_{i+1} = 1$.

Proposition 13.10 *The product of the hook lengths is given by the formula*

$$\prod h(i, j) = \frac{\ell_1! \ell_2! \cdots \ell_k!}{\prod_{i<j}(\ell_i - \ell_j)}.$$

Sketch of Proof Below we've entered the hook lengths for the boxes in the first row of the diagram λ:

Notice that their product is "almost" $\ell_1! = 11!$, but that some factors are missing. Check that the expression

$$\prod_{j=2}^{k}(\ell_1 - \ell_j)$$

accounts for these missing factors. Thus the product of the hook lengths from the first row is given by

$$\frac{\ell_1!}{\prod_j(\ell_1 - \ell_j)},$$

and the result follows by continuing down the rows. By now the serious reader should know to work through more examples.   □

We next establish a polynomial identity.

Lemma 13.11 Let $\Delta(x_1, \ldots, x_k)$ denote the Vandermonde polynomial (Exercise 5.12). Then

$$\sum_{i=1}^{k} x_i \Delta(x_1, \ldots, x_i + y, \ldots, x_k) = \left(x_1 + \cdots + x_k + \binom{k}{2}y \right) \Delta(x_1, \ldots, x_k),$$

where $\binom{k}{2}$ is a binomial coefficient.

Exercise* 13.12 Prove Lemma 13.11. Hints:

(1) Since the left-hand side is alternating in the variables $x_1, \ldots, x_k$, it is divisible by the Vandermonde polynomial.
(2) The quotient must be a homogeneous linear polynomial in the variables $x_1, \ldots, x_k$, and y. The coefficients for each of the x_i must be equal to 1, as can be seen by expanding both sides enough to compare coefficients for the leading term $x_1 x_2 x_3^2 \cdots x_k^{k-1}$.
(3) The identity is certainly true for $y = 0$. To determine the coefficient for y, substitute appropriate values for the x_i and y so that both sides are non-zero.

Proof of Proposition 13.4 Start by defining

$$F(\ell_1, \ldots, \ell_k) := \frac{n!}{\prod h(i, j)} = \frac{n! \prod_{i<j}(\ell_i - \ell_j)}{\ell_1! \ell_2! \cdots \ell_k!},$$

the last equality by Proposition 13.10. Then the recursive formula given by Eq. 13.6 translates as

Equation 13.13

$$F(\ell_1, \ldots, \ell_k) = \sum_{i=1}^{k} F(\ell_1, \ldots, \ell_i - 1, \ldots, \ell_k),$$

where we set $F(\ell_1, \ldots, \ell_i - 1, \ldots, \ell_k) = 0$ if $\lambda^- \not\lessdot \lambda$.

So we are done if we can verify Eq. 13.13, that is, that the right side of Eq. 13.5 satisfies the same recursion formula as Eq. 13.6.

As in Exercise 5.12, we define

$$\Delta(\ell_1, \ldots, \ell_k) := \prod_{1 \leq i < j \leq k} (\ell_i - \ell_j).$$

Using this notation, and combining terms using common denominators, yields

$$\sum_{\lambda^- \lessdot \lambda} f^{\lambda^-} = \sum_{i=1}^{k} \frac{(n-1)! \, \Delta(\ell_1, \ldots, \ell_i - 1, \ldots, \ell_k)}{\ell_1! \cdots (\ell_i - 1)! \cdots \ell_k!}$$

$$= \frac{(n-1)! \sum \ell_i \, \Delta(\ell_1, \ldots, \ell_i - 1, \ldots, \ell_k)}{\ell_1! \cdots \ell_k!}.$$

Applying Lemma 13.11 to part of the numerator, with $x_i = \ell_i$ and $y = -1$, gives

$$\sum_{i=1}^{k} \ell_i \Delta(\ell_1, \ldots, \ell_i - 1, \ldots, \ell_k) = [\ell_1 + \ell_2 + \cdots + \ell_k - \binom{k}{2}] \Delta(\ell_1, \ldots, \ell_k).$$

Now observe that

$$\begin{aligned}
\ell_1 + \ell_2 + \cdots + \ell_k &= (\lambda_1 + k - 1) + (\lambda_2 + k - 2) + \cdots + (\lambda_k) \\
&= [\lambda_1 + \cdots + \lambda_k] + [(k-1) + (k-2) + \cdots + 1 + 0] \\
&= n + k(k-1)/2 \\
&= n + \binom{k}{2}.
\end{aligned}$$

Putting all this together:

$$
\begin{aligned}
f^\lambda &= \sum_{\lambda^- \prec \lambda} f^{\lambda^-} \\
&= \sum_{i=1}^{k} \frac{(n-1)!\, \triangle(\ell_1, \ldots, \ell_i - 1, \ldots, \ell_k)}{\ell_1! \cdots (\ell_i - 1)! \cdots \ell_k!} \\
&= \frac{(n-1)! \sum \ell_i \, \triangle(\ell_1, \ldots, \ell_i - 1, \ldots, \ell_k)}{\ell_1! \cdots \ell_k!} \\
&= \frac{(n-1)!\,[\ell_1 + \ell_2 + \cdots + \ell_k - \binom{k}{2}] \triangle(\ell_1, \ldots, \ell_k)}{\ell_1! \cdots \ell_k!} \\
&= \frac{(n-1)!\,[n + \binom{k}{2} - \binom{k}{2}] \triangle(\ell_1, \ldots, \ell_k)}{\ell_1! \cdots \ell_k!} \\
&= \frac{n!\, \triangle(\ell_1, \ldots, \ell_k)}{\ell_1! \cdots \ell_k!} \\
&= \frac{n!}{\prod h(i,j)}.
\end{aligned}
$$

$\square$

Exercise 13.14 Verify that

$$
f^\lambda = \sum_{\lambda^- \prec \lambda} f^{\lambda^-} = \sum_{i=1}^{k} \frac{(n-1)!\, \triangle(\ell_1, \ldots, \ell_i - 1, \ldots, \ell_k)}{\ell_1! \cdots (\ell_i - 1)! \cdots \ell_k!}
$$

for a few simple examples.

13.2 Branching Relations

We now use the above dimension calculation in the determination of how the irreducible representations of S_n decompose when restricted to S_{n-1}. Not surprisingly, there are a number of different proofs of the following result. This one follows [Si], Section VI.4, and uses the machinery of Young symmetrizers in the group algebra $\mathbb{C}[S_n]$. A later exercise will outline this result via Specht modules in $\mathbb{C}[\mathcal{T}^\lambda]$.

Recall from Chap. 8 that, given a standard Young tableau T with shape λ, the Young symmetrizer $\mathbf{E}_T$ generates an irreducible representation of S_n in $\mathbb{C}[S_n]$; namely,

$$
V^\lambda := \mathbb{C}[S_n]\mathbf{E}_T = \{a\mathbf{E}_T \mid a \in \mathbb{C}[S_n]\}.
$$

We wish to restrict this representation to S_{n-1}, but what do we mean by "restriction" in the context of the group algebra?

Definition 13.15 Let G be a finite group, and let H be a subgroup of G. If

$$a = \sum_{g_i \in G} a_i g_i \in \mathbb{C}[G],$$

then the *restriction of* a *to* H is given by

$$a|_H = \sum_{\substack{g_i \in G \\ a_i = 0 \text{ if } g_i \notin H}} a_i g_i \in \mathbb{C}[H].$$

In other words, we just omit those terms in a that are not in $\mathbb{C}[H]$. For example, if

$$a = (1) + 2(1,2) - 4(1,3) + 2(1,3,2) \in \mathbb{C}[S_3],$$

then

$$a|_{S_2} = (1) + 2(1,2) \in \mathbb{C}[S_2].$$

We start with some exercises that illustrate the main ideas.

Exercise* 13.16 Let

$$T = \begin{array}{|c|c|} \hline 1 & 2 \\ \hline 3 \\ \cline{1-1} \end{array}$$

be a tableau with shape $\lambda = (2, 1, 0) \vdash 3$.

(1) What is the dimension of $V^{(2,1,0)}$?
(2) Write out $\mathbf{E}_T$.
(3) Observe that the elements $\mathbf{E}_T$ and $(1,3).\mathbf{E}_T$ are linearly independent vectors, and so are a basis for the irreducible representation $V^{(2,1,0)}$ in $\mathbb{C}[S_3]$.
(4) Restrict $\mathbf{E}_T$ and $(1,3).\mathbf{E}_T$ to S_2 and describe the subspace of $\mathbb{C}[S_2]$ spanned by these vectors.
(5) How does the space in part (4) decompose into irreducible representations of S_2?

Exercise* 13.17 Let

$$T_1 = \begin{array}{|c|c|} \hline 1 & 2 \\ \hline 3 \\ \cline{1-1} \end{array} \quad \text{and} \quad T_2 = \begin{array}{|c|c|} \hline 1 & 3 \\ \hline 2 \\ \cline{1-1} \end{array}$$

be the two standard Young tableaux with shape $(2, 1)$. Let T_1^- and T_2^- be the tableaux obtained by removing the box containing the entry 3 from the tableaux T_1 and T_2 respectively.

Recall that the Young symmetrizers $\mathbf{E}_{T_1}$ and $\mathbf{E}_{T_2}$ generate two copies of the irreducible representation $V^{(2,1,0)}$ of S_3 in $\mathbb{C}[S_3]$. Write out $\mathbf{E}_{T_1^-}$ and $\mathbf{E}_{T_2^-}$, and check that each symmetrizer generates an irreducible representation of S_2. What are they, what are their dimensions and how does this relate to the dimension of $V^{(2,1,0)}$? You should also try this with the representation $V^{(3,1,0,0)}$ of S_4.

We now proceed to the general case.

Proposition 13.18 *Let $\lambda^- \prec \lambda$, let $V^\lambda \in \mathbb{C}[S_n]$ be an irreducible representation of S_n labeled by λ, and let $V^{\lambda^-} \in \mathbb{C}[S_{n-1}]$ be an irreducible representation of S_{n-1} labeled by λ^-. Then*

$$\text{Res}^{S_n}_{S_{n-1}} V^\lambda = \bigoplus_{\lambda^- \prec \lambda} V^{\lambda^-}.$$

Proof For $\lambda^- \prec \lambda$, let T be a tableau with shape λ that assigns n to the box that is removed by going from λ to λ^-, and denote the corresponding tableau by T^-. A quick check verifies that, for the vertical and horizontal subgroups, we have

$$V_{T^-} = V_T \cap S_{n-1}, \quad \text{and} \quad H_{T^-} = H_T \cap S_{n-1}.$$

Also check that if $hv \in S_{n-1}$ for $h \in H_T$ and $v \in V_T$, then both h and v must lie in S_{n-1}, since if the entry n is first moved by a column permutation, it cannot be returned by a row permutation. Therefore,

$$\mathbf{E}_T|_{S_{n-1}} = \mathbf{E}_{T^-}.$$

Now

$$V^{\lambda^-} = \mathbb{C}[S_{n-1}]\mathbf{E}_{T^-}$$

is an irreducible representation of S_{n-1}, and since $\mathbf{E}_{T^-}$ is contained in both V^{λ^-} and $V^\lambda|_{S_{n-1}}$, we have $V^{\lambda^-} \cap V^\lambda|_{S_{n-1}} \neq \{0\}$. By Exercise 2.35, V^{λ^-} is an S_{n-1}-irreducible subrepresentation of $V^\lambda|_{S_{n-1}}$. Since this is true for each $\lambda^- \prec \lambda$, we have

$$\operatorname{Res}_{S_{n-1}}^{S_n} V^\lambda \supset \bigoplus_{\lambda^- \prec \lambda} V^{\lambda^-},$$

the sum being direct by the irreducibility of each V^{λ^-}, and by Exercise 2.35 again. By Eq. 13.6, the dimensions of both sides are equal, which completes the proof. □

Remark 13.19 Note that all of the irreducible representations comprising $\operatorname{Res}_{S_{n-1}}^{S_n} V^\lambda$ appear at most once. Such a decomposition is said to be *multiplicity-free*, and there is a considerable body of research concerning multiplicity-free representations of various groups.

Exercise* 13.20 Here is another way to look at restricted representations. See [Sa], Section 2.8 for an in-depth discussion. Let $\lambda = (3, 2) \vdash 5$, and consider the standard Young tableaux

$$T_1 = \begin{array}{|c|c|c|} \hline 1 & 3 & 5 \\ \hline 2 & 4 \\ \hline \end{array}, \quad T_2 = \begin{array}{|c|c|c|} \hline 1 & 2 & 5 \\ \hline 3 & 4 \\ \hline \end{array},$$

$$T_3 = \begin{array}{|c|c|c|} \hline 1 & 2 & 3 \\ \hline 4 & 5 \\ \hline \end{array}, \quad T_4 = \begin{array}{|c|c|c|} \hline 1 & 2 & 4 \\ \hline 3 & 5 \\ \hline \end{array}, \quad \text{and} \quad T_5 = \begin{array}{|c|c|c|} \hline 1 & 3 & 4 \\ \hline 2 & 5 \\ \hline \end{array}.$$

For each i, let $\mathcal{V}_{T_i}$ be the column anti-symmetrizer, and recall from Chap. 11 that the standard polytabloids $\mathbf{e}_{T_i} = \mathcal{V}_{T_i}.\{T_i\}$ form a basis for the Specht module $\mathcal{S}^{(3,2)}$ in $\mathbb{C}[\mathcal{T}^{(3,2)}]$.

(1) Let T_i^- denote the tableau T_i with the box containing the entry 5 removed, and let $\mathcal{V}_{T_i}^-$ denote the restriction of $\mathcal{V}_{T_i} \in \mathbb{C}[S_5]$ to $\mathbb{C}[S_4]$. Show that $\mathcal{V}_{T_i}^- = \mathcal{V}_{T_i^-}$.

(2) Let $\mathbf{e}_{T_i}^- = \mathcal{V}_{T_i}^-.\{T_i\}$. Show that $\mathbf{e}_{T_1}^-$ and $\mathbf{e}_{T_2}^-$ span a copy of the Specht module $\mathcal{S}^{(2,2)}$ contained in $\mathcal{S}^{(3,2)}$, and that $\mathbf{e}_{T_3}^-$, $\mathbf{e}_{T_4}^-$, and $\mathbf{e}_{T_5}^-$ span a copy of the Specht module $\mathcal{S}^{(3,1)}$ contained in $\mathcal{S}^{(3,2)}$. Conclude that $\operatorname{Res}_{S_4}^{S_5} \mathcal{S}^{(3,2)} \cong \mathcal{S}^{(2,2)} \oplus \mathcal{S}^{(3,1)}$.

We now proceed to the decomposition of induced representations. Given a Young diagram $\lambda \vdash n$, an *inner corner* is a location where a box can be added both at the end of a row and also at the bottom of a column (one of these conditions may be vacuously satisfied). We denote by λ^+ any Young diagram that can be obtained from λ by adding one box at an inner corner, and it follows from the definition that $\lambda^+ \vdash n + 1$ is a legitimate Young diagram. We use the notation $\lambda^+ \succ \lambda$ when λ^+

can be obtained from λ by adding a box at an inner corner. Below we've bulleted these inner corners,

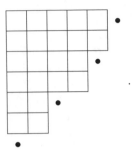

Now the branching rules say that the irreducible representation V^{λ^-} appears in Res V^λ exactly once for each $\lambda^- \prec \lambda$. Boosting this up from S_n to S_{n+1} says that each V^λ appears exactly once in Res V^{λ^+} for each $\lambda \prec \lambda^+$. Frobenius reciprocity then says that V^{λ^+} appears in Ind V^λ exactly once for each $\lambda \prec \lambda^+$. Applying this to Proposition 13.18 yields the following corollary.

Corollary 13.21

$$\mathrm{Ind}_{S_n}^{S_{n+1}} V^\lambda = \bigoplus_{\lambda^+ \succ \lambda} V^{\lambda^+}.$$

Exercise 13.22 Decompose $\mathrm{Ind}_{S_n}^{S_{n+1}} V^\lambda$ for several examples of n and λ.

A *Hasse diagram* is a graphical rendering of a partially ordered set. The figure below is the Hasse diagram (also called the *Young lattice*) that depicts the branching relations for the irreducible representations of the symmetric groups. Move down along the edges for restriction, and move up along the edges for induction.

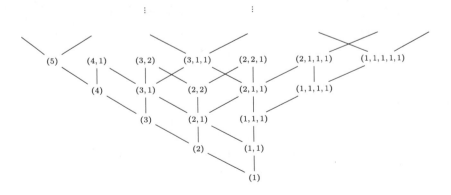

For example,

$$\text{Res}_{S_4}^{S_5}(3, 1, 1) = (3, 1) \oplus (2, 1, 1), \quad \textbf{and} \quad \text{Ind}_{S_3}^{S_4}(2, 1) = (3, 1) \oplus (2, 2) \oplus (2, 1, 1).$$

Since restriction and induction are transitive,

$$\text{Res}_{S_3}^{S_5}(3, 1, 1) = (3, 0, 0) \oplus 2(2, 1) \oplus (1, 1, 1),$$

and

$$\text{Ind}_{S_3}^{S_5}(2, 1) = (4, 1) \oplus 2(3, 2) \oplus (3, 1, 1) \oplus (2, 2, 1) \oplus (2, 1, 1, 1).$$

Exercise 13.23 Reproduce the Young lattice by labeling the nodes with Young diagrams instead of partitions. Extend it upward a few more levels. Hint: It's easier to do this on a large white or black board.

Observe that we move up along the edges of the Young lattice by adding a box to an inner corner. Also observe that the sequence of adding boxes, for example the sequence

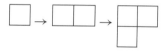

corresponds to the standard Young tableau

$$\begin{array}{|c|c|} \hline 1 & 2 \\ \hline 3 \\ \cline{1-1} \end{array}.$$

Conclude that the dimension of the irreducible representation $\mathcal{S}^\lambda$ is equal to the number of distinct paths from the trivial diagram to the diagram for λ in the Young lattice.

Exercise* 13.24 We'll end with a fun/interesting exercise. Let F denote the vector space of differentiable functions on $\mathbb{R}$. Let x denote the linear map obtained by multiplying some $f \in F$ by the variable x, and let d_x denote the linear map obtained by taking the derivative of f with respect to x. Show that for the commutator product (Remark 6.1) we have

$$[d_x, x] := d_x x - x d_x = I_F,$$

where I_F is the identity map on F.

Let R denote the vector space spanned by the irreducible representations of S_n. Let $\uparrow$ denote the operation of induction from S_n to S_{n+1}, and let $\downarrow$ denote the operation of restriction from S_n to S_{n-1}, both operations acting linearly on the representations of S_n for all n. Work a few examples to see that

$$[\downarrow, \uparrow] := \downarrow\uparrow - \uparrow\downarrow = I_R,$$

Where I_R is the identity on R. Ambitious readers can prove this for the general case using Proposition 13.18 and its corollary.

13.3   Hints and Additional Comments

Exercise 13.1 If a tableau with shape λ is standard, then the entries are increasing along the rows and down the columns. Therefore the corresponding entries in the conjugate tableau with shape λ' are increasing down the columns and along the rows.

Exercise 13.12 The first two hints establish that

Equation 13.25

$$\sum_{i=1}^{k} x_i \triangle(x_1, \ldots, x_i + y, \ldots, x_k) = (x_1 + \cdots + x_k + cy)\triangle(x_1, \ldots, x_k)$$

for some constant c.

The key to determining the value of c is to make an efficient substitution for the variables. Set $x_i = k - i$, set $y = 1$, and consider the right-hand side of Eq. 13.11. Check that, after this substitution, we have

Equation 13.26

$$\triangle(x_1, \ldots, x_k) = \prod_{1 \le i < j \le k} (x_i - x_j) = \prod_{1 \le i < j \le k} (j - i) = (-1)^q \prod_{i=1}^{k-1} i!,$$

where $q = 1 + 2 + \cdots + (k - 1)$ is the number of factors in the product.

For the linear factor on the right-hand side of Eq. 13.25, we have

$$x_1 + \cdots + x_k + cy = (k-1) + (k-2) + \cdots + 2 + 1 + c = \frac{k(k-1)}{2} + c = \binom{k}{2} + c.$$

For the left-hand side of Eq. 13.25, observe that in the expression

$$\triangle(x_1, \ldots, x_i + t, \ldots, x_k)$$

we have factors of the form $x_{i+1} + y - x_i = k - (i+1) + 1 - (k-i)$ that evaluate to zero upon substitution. Thus each term in the sum on the left-hand side is equal to zero except for the first term, namely, $x_1 \triangle(x_1 + y, x_2, \ldots, x_k)$. Substituting and evaluating this first term yields

Equation 13.27

$$x_1 \triangle(x_1 + t, \ldots, x_k) = (-1)^q k! \frac{(k-1)!}{k-2} \frac{(k-2)!}{k-3} \cdots 2! = (-1)^q \frac{\prod_{i=1}^k i!}{(k-2)!}.$$

So Eq. 13.25 becomes

$$(-1)^q \frac{\prod_{i=1}^k i!}{(k-2)!} = \left[\binom{k}{2} + c \right] (-1)^q \prod_{i=1}^{k-1} i!.$$

Some cancellation gives

$$\frac{k!}{(k-2)!} = \left[\binom{k}{2} + c \right], \quad \text{so that} \quad k(k-1) = \frac{k(k-1)}{2} + c,$$

and hence $c = \frac{k(k-1)}{2} = \binom{k}{2}$.

Note: To see the right-hand side of Eq. 13.26, write out $\triangle(x_1, \ldots, x_k)$ as

$$
\begin{aligned}
\triangle(x_1, \ldots, x_k) &= \prod_{i<j}(x_j - x_i) \\
&= (x_k - x_{k-1})(x_k - x_{k-2}) \cdots (x_k - x_1) \\
&\quad \times (x_{k-1} - x_{k-2})(x_{k-1} - x_{k-3}) \cdots (x_{k-1} - x_1) \\
&\quad \vdots \\
&\quad \times (x_2 - x_1),
\end{aligned}
$$

and then make the substitution $x_i = k - 1$. A slight modification to this viewpoint establishes Eq. 13.27.

Exercise 13.16

$$\mathbf{E}_T = (1) + (1, 2) - (1, 3) - (1, 3, 2),$$

and

$$(1, 3).\mathbf{E}_T = -(1) + (1, 3) - (2, 3) + (1, 2, 3).$$

Upon restriction,

$$\mathbf{E}_T|_{S_2} = (1) + (1, 2),$$

and

$$(1, 3).\mathbf{E}_T|_{S_2} = -(1).$$

These span the space

$$\langle\!\langle \mathbf{E}_T|_{S_2}, (1, 3).\mathbf{E}_T|_{S_2} \rangle\!\rangle = \langle\!\langle (1) + (1, 2), -(1) \rangle\!\rangle = \langle\!\langle (1), (1, 2) \rangle\!\rangle = \mathbb{C}[S_2],$$

which decomposes as $V^{(2,0)} \oplus V^{(1,1)}$.

Exercise 13.17 Let $\lambda = (2, 1, 0) \vdash 3$, and let

$$T_1 = \begin{array}{|c|c|} \hline 1 & 2 \\ \hline 3 \\ \cline{1-1} \end{array}, \quad \text{so that} \quad T_1^- = \begin{array}{|c|c|} \hline 1 & 2 \\ \hline \end{array}.$$

Then

$$\mathbf{E}_{T_1^-} = (\mathbf{1}) + (\mathbf{1, 2}),$$

which generates the trivial representation $V^{(2,0)}$ of S_2. Note that

$$\mathbf{E}_{T_1} = (\mathbf{1}) + (\mathbf{1, 2}) - (\mathbf{1, 3}) - (\mathbf{1, 3, 2}), \quad \text{and that} \quad \mathbf{E}_{T_1}|_{S_2} = \mathbf{E}_{T_1^-}.$$

Now let

$$T_2 = \begin{array}{|c|c|} \hline 1 & 3 \\ \hline 2 \\ \cline{1-1} \end{array}, \quad \text{so that} \quad T_2^- = \begin{array}{|c|} \hline 1 \\ \hline 2 \\ \hline \end{array}.$$

Then

$$\mathbf{E}_{T_2^-} = (\mathbf{1}) - (\mathbf{1, 2}),$$

which generates the alternating representation $V^{(1,1)}$ of S_2. Note that

$$\mathbf{E}_{T_2} = (\mathbf{1}) + (\mathbf{1, 3}) - (\mathbf{1, 2}) - (\mathbf{1, 2, 3}), \quad \text{and that} \quad \mathbf{E}_T|_{S_2} = \mathbf{E}_{T_2^-}.$$

It follows that

$$V^{(2,1,0)}|_{S_2} = V^{(2,0)} \oplus V^{(1,1)}.$$

Both $V^{(2,0)}$ and $V^{(1,1)}$ have dimension one, which sum to the dimension of $V^{(2,1,0)}$.

Exercise 13.20 The relevant observation here is that the map between the basis polytabloids, for example, the map between

$$\mathbf{e}_{T_3^-} = V_{T_3^-}.\{T_3^-\} = \;\begin{array}{|ccc|} \hline 1 & 2 & 3 \\ \hline 4 \\ \hline \end{array}\; - \;\begin{array}{|ccc|} \hline 2 & 3 & 4 \\ \hline 1 \\ \hline \end{array}\; \in \mathcal{S}^{(3,1)},$$

and

$$\mathbf{e}_{T_3^-} = V_{T_3^-}.\{T_3\} = \;\begin{array}{|ccc|} \hline 1 & 2 & 3 \\ \hline 4 & 5 \\ \hline \end{array}\; - \;\begin{array}{|ccc|} \hline 2 & 3 & 4 \\ \hline 1 & 5 \\ \hline \end{array}\; \in \mathbb{C}[\mathcal{T}^{(3,2)}],$$

intertwines the action of S_4, and so embeds a copy of $\mathcal{S}^{(3,1)}$ as a subrepresentation of $\mathcal{S}^{(3,2)}|_{S_4}$.

Exercise 13.24 Using the product rule for derivatives, for any $f \in F$ we have

$$[d_x, x](f) = d_x[x(f)] - xd_x(f) = f + xd_x(f) - xd_x(f) = f.$$

Thus, $[d_x, x] = I_F$.

This is a simplified version of the *canonical commutation relations* of quantum physics, where multiplication by x is the position operator, and d_x is the momentum operator. Since the commutator product is non-zero, the position and momentum operators don't commute, and therefore the operators are not simultaneously diagonalizable. In the yoga of quantum mechanics, this means that both position and momentum cannot by simultaneously measured, which is known as the *Heisenberg uncertainty principle*.

Induction and restriction on representations of the symmetric groups also obey this commutation relation. Here's a few computations to get you started.

$$\downarrow\uparrow (3, 1) = \downarrow \{(4, 1) + (3, 2) + (3, 1, 1)\} = (4) + (3, 1) + (3, 1) + (2, 2) + (3, 1) + (2, 1, 1).$$

and

$$\uparrow\downarrow (3, 1) = \uparrow \{(3) + (2, 1)\} = (4) + (3, 1) + (3, 1) + (2, 2) + (2, 1, 1),$$

The actions of restriction and induction are sometimes referred to as *Down operators* and *Up operators* in this context.

Bibliography

[AF] Anderson, F. W., & Fuller, K. R. (1974). *Rings and categories of Modules*. New York, Berlin, Heidelberg: Springer.

[C] Curtis, C. W. (1999). *Pioneers of representation theory: Frobenius, burnside, Schur, and Brauer*. Providence, RI: American Mathematical Society.

[D] Daugherty, Z., Eustis, A. K., Minton, G., & Orrison, M. E. (2009). Voting, the symmetric group, and representation theory. *The American Mathematical Monthly, 116*(8), 667–687.

[E] Etingof, P., Golberg, O., Hensel, S., Liu, T., Schwendner, A., Vaintrob, D., & Yudovina, E. (2011). *Introduction to representation theory. Student mathematical library* (Vol. 59). Providence, RI: American Mathematical Society.

[F] Fulton, W. (1997). *Young tableaux. London mathematical society student texts* (Vol. 35). Cambridge, New York, Melbourne: Cambridge University Press.

[FH] Fulton, W., & Harris, J. (1991). *Representation theory, a first course*. New York, Heidelberg, London, Paris, Tokyo, Hong Kong, Barcelona, Budapest: Springer.

[G] Gallian, J. A. (2010). *Contemporary abstract algebra* (7th ed.). Belmont, CA: Brooks/Cole.

[GN] Glass, K., & Ng, C.-K. (2004). A simple proof of the Hook length formula. *The American Mathematical Monthly, 111*(8), 700–703.

[H] Hartnet, K. (2020). The 'useless' perspective that transformed mathematics. *Quanta*, June 9, 2020. https://www.quantamagazine.org/the-useless-perspective-that-transformed-mathematics-20200609/

[HLV] Howe, R. M., Liu, H., & Vaughan, M. (2020). A proof of the equivalence between the polytabloid bases and Specht polynomials for irreducible representations of the symmetric group. *Ball State Mathematics Exchange, 14*(1) (Fall), 25–33.

[J] James, G. (1978). *The representation theory of the symmetric groups*. Berlin, Heidelberg-Berlin, Heidelberg: Springer.

[JK] James, G., & Kerber, A. (1985). *The representation theory of the symmetric group*. New York: Cambridge University Press.

[K] Knuth, D. E. (1998). *The art of computer programming III* (2nd ed.). Reading, MA: Addison-Wesley.

[L1] Lang, S. (1984). *Algebra* (2nd ed.). Menlo Park, CA: Addison-Wesley.

[L2] Lang, S. (2010). *Linear algebra* (3rd ed.). New York: Springer.

[M] Macdonald, I. G. (1995). *Symmetric functions and Hall polynomials* (2nd ed.) Cambridge, New York, Melbourne: Oxford University Press.

[MS] Mirman, R., & Schindler, S. (1977). The decomposition of tensor product representations of the symmetric group. *Journal of Mathematical Physics, 18,* 1678–1696.

[P] Peel, M. H. (1975). Specht modules and symmetric groups. *Journal of Algebra, 36,* 88–97.

[R] Rutherford, D. E. (2013). *Substitutional analysis*. New York: Dover Publications Inc.

© The Author(s), under exclusive license to Springer Nature Switzerland AG 2022
R. M. Howe, *An Invitation to Representation Theory*, SUMS Readings,
https://doi.org/10.1007/978-3-030-98025-2

[Sa] Sagan, B. E. (1991). *The symmetric group; representations, combinatorial algorithms, and symmetric functions* (2nd ed.). New York, Berlin, Heidelberg, Barcelona, Hong Kong, London, Paris, Singapore, Tokyo: Springer.

[Se] Serre, J.-P. (1977). *Linear representations of finite groups*. New York, Berlin, Heidelberg: Springer.

[Si] Simon, B. (1996). *Representations of finite and compact groups*. New York: American Mathematical Society.

[Sm] Smith, K. E. (2014). *Bases for Infinite Dimensional Vector Spaces*. http://www.math.lsa. umich.edu/~kesmith/infinite.pdf

[Sp] Specht, W. (1935). Die irreduziblen Darstellungen der symmetrischen Gruppe. *Mathematische Zeitschrift, 39*, 696–711.

[St] Sternberg, S. (1994). *Group Theory and Physics*. Cambridge, New York, Melbourne: Cambridge University Press.

[W] Weyl, H. (1950). *The theory of groups and quantum mechanics*. New York: Dover Publications Inc.

Index

© The Author(s), under exclusive license to Springer Nature Switzerland AG 2022
R. M. Howe, *An Invitation to Representation Theory*, SUMS Readings,
https://doi.org/10.1007/978-3-030-98025-2

Printed in the United States
by Baker & Taylor Publisher Services